ELEMENTS OF TIDAL-ELECTRIC ENGINEERING

IEEE Press
445 Hoes Lane
Piscataway, NJ 08854

ELEMENTS OF TIDAL-ELECTRIC ENGINEERING

ROBERT H. CLARK

IEEE

IEEE PRESS

BICENTENNIAL 1807 WILEY 2007

WILEY-INTERSCIENCE

A JOHN WILEY & SONS, INC., PUBLICATION

Unless indicated otherwise, all illustrations are owned and copyrighted by the author.

Published by John Wiley & Sons, Inc. Published simultaneously in Canada.

Library of Congress Cataloging-in-Publication Data is available.

ISBN 978-0-470-10709-6

Printed in the United States of America

10 9 8 7 6 5 4 3 2 1

With love
to my wife,
Ruth

In memory of
Robert H. Clark
1921–2007

CONTENTS

PREFACE

We think of tidal power as a relatively new source of energy, yet it has been available since the oceans existed and our satellite, the Moon, has circled planet Earth. For at least a thousand years, our ancestors have harnessed and put a very minute portion of this natural energy to good use. Judging by the volume of literature dating back several centuries, the practical-minded and the dreamers of the time have contemplated ways of extracting more of the enormous amounts of energy available in the tides that can be observed rising and falling relentlessly in many coastal estuaries. In fact, during the past two centuries, several hundred patents have been filed to protect ingenious devices and concepts for extracting this energy.

Since the availability of tidal energy for electrical conversion is limited to areas where relatively large tidal ranges occur, there is not the same widespread interest in exploiting this resource as, for example, there is in hydroelectric developments, which make up about 23 percent of the global installed, electrical generating capacity. Tidal power technology is relatively mature and, at the present time, there are three commercially operating tidal-electric stations with capacities greater than 3 MW—one each in France, China, and Canada.

Although river hydropower and tidal power have many characteristics in common, there is a fundamental difference in the forces activating these two energy sources, viz. the downward-acting terrestrial gravitation causing water to flow on Earth's surface and the upward-acting lunar gravitation causing water to oscillate in estuaries. There is a tendency to ignore this basic dissimilarity and to treat tidal-electric developments as low-head hydroelectric developments.

The treatment of the subject matter in this text is designed to bridge the gap in the exploitation of hydro and tidal energy resources. The objective is to present information on the major aspects to be dealt with in a feasibility investigation. Following a brief overview of the history of tidal energy exploitation from the era of tidal mills, enabling the conversion to mechanical energy, to the latter half of the twentieth century in which conversion to electrical energy has been successfully accomplished (Chapter 1), the phenomenon of the tides is reviewed (Chapter 2).

Chapter 3 deals with the estimation of site potential and reviews the major factors to be considered in selecting a site for preliminary assessments. In view of the complexity and interrelationship of the many activities involved in a tidal power feasibility investigation, Chapter 4 deals with the setting up of an appropriate organization to carry out the studies.

The types and schemes to exploit tidal energy are reviewed in Chapter 5. Chapter 6 outlines the types of physical data of an estuary needed to make a reliable assessment of the feasibility of a potential site. Chapter 7 discusses the essential contribution provided by hydraulic and mathematical models of a selected estuary in order to predict not only its response to physical changes but also the effects of operation of the proposed tidal-electric plant.

The civil works required for a tidal power development, including the relative merits of "floated-in" construction and construction "in-the-dry" for a potential site, and the associated tidal generating equipment are reviewed in Chapters 8 and 9, respectively. Optimization procedures to select the size and characteritics of turbogenerating machinery to suit specific site conditions and output requirements and to determine the number of sluice and generating units for maximum energy production at the lowest overall cost per unit of output are discussed in Chapter 10. Chapter 11 deals with the problem of integrating the output of a tidal plant, which is in phase with the lunar cycle, with that from other electrical generating sources of a power system in order to meet efficiently the system load demands. The latter are in phase with the solar cycle.

An approach to the economic evaluation of a tidal power plant is outlined in Chapter 12 together with a brief review of some of the factors influencing the main parameters in an economic analysis that may involve uncertainties. Chapter 13 discusses potentially significant social and regional issues that may result from the construction of a tidal power development, while Chapter 14 reviews potential interactions of the construction and operation of a tidal plant on elements of the environment. The construction and operation of existing, commercially operating, tidal-electric plants are reviewed in Chapter 15, and a brief review of potential sites that have been the subject of preliminary and prefeasibility investigations during the past half-century is provided in Chapter 16.

The aim of the text, which has been based on notes designed and developed by the author to support a series of lectures he presented on tidal power under UNDP sponsorship to the staff of the Central Water and Power Research

Station at Pune, India, in early 1982, is to collate and present information on major aspects to be dealt with in an investigation of a tidal power development. The text is not intended to be, nor does it purport to be, an exhaustive treament of the subject but, rather, it is designed to provide an introduction to the subject of tidal-electric engineering. The original notes have been expanded to include additional material presented during the lectures and to update the status of the subject as a result of investigations undertaken since that time. Obviously, some aspects of the text may require elaboration that can be undertaken by the reader on the basis of the references provided.

ROBERT H. CLARK

Calgary, Alberta, Canada
December 2006

ACKNOWLEDGMENTS

Tidal-electric engineering requires the expertise of many fields including system planning, coastal engineering, hydroelectric engineering, mechanical and electrical engineering, hydraulic engineering, and environmental, as well as economic and financial skills. The depth and breadth of knowledge required from each of these disciplines depend upon the type of investigation undertaken. Appropriate inputs by many specialists for specific aspects of these fields is essential to meet successfully the particular objective of an investigation.

Because of the myriad of specialist subjects involved in an investigation of a tidal-electric development, whether it be of the reconnaissance type or a definitive feasibility study, it would be presumptuous for one person to write *authoritatively* on all aspects. The material for these notes has been collected from many sources: governmental reports on tidal power assessments and prefeasibility studies, engineering and scientific journals, results of symposia, workshops and conferences on energy and tidal power, and the like. The author gratefully acknowledges the works of expert and specialist colleagues that form the prime ingredients for this text. In particular, I would like to thank J. Gavin Warnock, P.Eng., Consulting Engineer, and Norman Crookshank, P.Eng., Senior Project Engineer, Canadian Hydraulics Centre. I am very much obliged to George C. Baker, P.Eng., Executive Director of the former Tidal Power Corporation, Halifax, for his professional advice and editorial suggestions.

R. H. C.

CHAPTER 1

HISTORICAL DEVELOPMENT

Theoretically, any tide anywhere can produce power in some quantity from the energy otherwise dissipated during its ebb stage, but there are relatively few locations in the world where topographical conditions are such as (1) to cause the incoming tide to rise to an unusual or abnormal height through being forced to its culminating level in a long gradually contracting channel and (2) to permit a stable barrier to be placed in the path of the tide, provided with openings to allow the tidal flow to pass through to its culminating level and thereafter capable of being closed to hold this level as the tide recedes on the seaward side of the barrier. Under such a condition, it follows that, as the tide recedes seaward, or *ebbs*, a gradually increasing head differential is created between the falling seaward levels and the fixed level held above and within the barrier, designated the *tidal basin*. When this head differential has reached some predetermined value the potential energy thereby created through the agency of the tidal flow impounded by the barrier can be converted into controllable mechanical or electrical power. Thus, in addition to a sufficiently high tidal range, at least 5 m as a rule-of-thumb for construction of an economically feasible tidal power plant, the site should also include a natural bay with an adequate area and volume and be so situated that the operation of the plant will not significantly reduce the tidal range.

There are two limitations associated with the production of this tidal power. First, the power potential will vary during the approximately 12-h cycle of ebb and flow. Second, a similarly recurring cessation of all power production

Elements of Tidal-Electric Engineering. By Robert H. Clark

during a period that begins when the rising tide, coupled with the falling head in the tidal basin, has eliminated the operable head, and ends when the tide has receded to a level resulting in the predetermined minimum operating head at which the turbine can function and with the tidal basin again full to high slack level. In principle, therefore, a simple tidal development can only produce power intermittently within the daily tidal cycles.

The energy in the ebb and flow of the tides, particularly in those areas of the world where the tides exhibit a relatively large range, has long been recognized as a naturally repetitive source of energy. In fact, this source was tapped more than a millenium ago to carry out some of the more tedious tasks, such as the grinding of grains.

According to historical records, inhabitants of the coastlines, particularly those bordering the North Atlantic, installed simple tidal mills on the shores of Gaul, Andalusia, and England, at least as early as the Middle Ages, to aid them in their daily work. It is likely that tidal energy also tapped in other countries, perhaps even earlier.

The tidal mill is a version of the conventional water mill but required just a little extra skill and care to operate it. The dam, closing off the small bay or mill pond from the sea, was equipped with sluice gates, or *flap* gates. These opened automatically with the tidal current during the *flood* and closed during the *ebb* to trap the water in the mill pond. From there it was directed to drive a large-diameter, wooden paddle wheel and thence back to the sea. A schematic arrangement is shown in Fig. 1.1.

These ancient mills used the ebb tide for the generating cycle, and the engineering skill of those who built them is to be admired. The great driving wheels, made of wood and subsequently of cast iron, mostly with replaceable wooden teeth, matched exactly the smaller wheels that they operated. Many mills were automated to the extent that they could be operated by one person who had readily available a series of ropes and pulleys by which any part of the mill's machinery could be set in motion. Since the tidal cycle and solar cycle are out of phase in most locations around the world, the miller had to mill when the tide was right whether at midnight or midday—an early example of flexible working hours!

The earliest records of tidal mills date from about the eleventh century. For the next several centuries, many tidal mills were reported to have been in operation along the Atlantic coast of Europe, mostly in the British Isles,

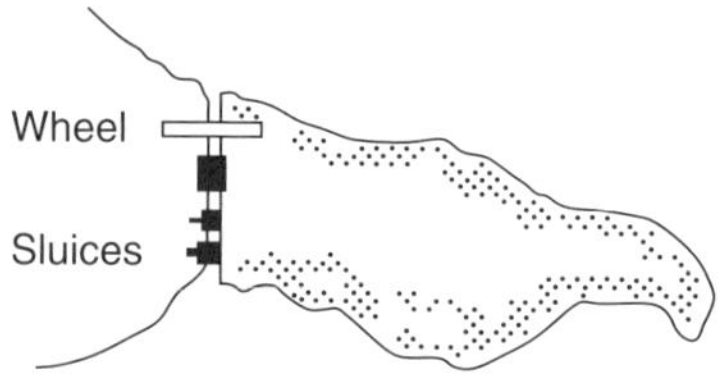

Figure 1.1 Typical tidal mill setup.

France, and Spain. One such installation in the Deben Estuary (England) was mentioned as early as 1170 in the records of the Parish of Woodbridge. The present building there (Fig. 1.2) dates from the eighteenth century, and the mill was in operation until 1957 when the main wheel shaft broke. Since then, the mill pond has been converted into a yacht marina and the mill restored as a museum piece.

Two tidal mills were still in operation in the Rance Estuary (Fig. 1.3) when Electricité de France began work leading to the construction of the world's first large, modern tidal-electric power plant. Tidal energy was tapped as early as 1617 in North America. Slade's Mill in Chelsea, Massachusetts, built about 1734 to grind spices and developed about 35 kW. Prior to 1800, at least two small, single-basin tidal mills were in existence in Passamaquoddy Bay. Also according to Bernstein (1965), there were tidal mills in Russia as early as the eighteenth century.

Besides the use of the simple water wheel turned by the tide and connected by belts or gears to grinding or *mill* stones, relatively primitive devices were developed to harness the tidal energy for other uses, such as for pumping. Even in 1824, part of the water supply of London was still provided by 6-m diameter water wheels installed in 1580 under the arches of London Bridge. Some of these structures for transforming tidal energy were of impressive size.

Figure 1.2 Tidal mill on the Deben Estuary. (Courtesy Woodbridge Tide Mill Trust.)

Figure 1.3 Tidal mill in the Rance Estuary. (Courtesy of Electricité de France.)

A tidal mill built at the beginning of the eighteenth century on the shores of Rhode Island, United States, had 3.36-m diameter wheels that were 7.92 m in width and weighed about 20 tons.

Many ingenious methods, in addition to the water wheel, were devised to use the potential energy of the tides or the kinetic energy of the tidal currents, or combinations of both, by means of such devices as lifted platforms or weights and air compressors. At least 300 patents have been registered during the past 100 years for different technical systems to extract the energy of the tides. However, all of these earlier methods had one common characteristic—they were devised to transform the energy of the tides into mechanical energy for local consumption.

An interesting survey, with an extensive bibliography, of the development and use of tidal power from classical times up to the completion of La Rance (France) tidal-electric plant in 1967 is presented by Charlier (1969).

The early tidal developments could extract only an infinitesimally small fraction of the potentially available energy, producing, perhaps, the equivalent of up to 75 kW for use at the site. Such micro amounts of power served the needs of the community before the advent of the electric motor and long-distance power transmission.

Efforts to rationalize tidal power technology were undertaken centuries ago. A handbook for the construction of tidal mills was written by an Italian,

Mariano, in the fifteenth century. At the beginning of the eighteenth century, a Frenchman, Belidor, considered the "quality" of tidal energy. In his treatise on hydraulic architecture, he stated principles for multibasin operation to achieve a continuous output of energy.

The advent of the Industrial Revolution introduced insatiable demands for power and energy on a large scale. The exploitation of tidal energy fell into disuse because hydroelectric development of rivers and fossil-fired power plants offered easier technological access to power generation on an industrial scale. The advances in technology for the development of these latter sources of power resulted in neglect of tidal power technology. The small tidal mills, not being able to meet the competition, disappeared from the scene—monuments of a technology that had outlived its day.

During the past four decades there have been a number of major advances in technology, notably in hydraulic turbogenerating units, in marine construction, and in the mathematical understanding of tidal cycle variations. It is now clear that the major technological problems formerly associated with large-scale tidal power developments have been resolved.

It has only been since the 1960s that the construction of modern tidal-electric stations has been attempted. On November 26, 1966, the first large-scale, modern, tidal-electric plant went into operation in La Rance Estuary near St. Malo, France. Less than a year later, the plant was fully equiped with 24–10,000 kW turbogenerators and has been operating successfully since that time as an integral part of the electrical utility system of Electricité de France.

During the latter part of 1968, an experimental tidal power plant commenced operation at the mouth of Kislaya Gulf near Murmansk, Russia, on the coast of the Barents Sea. The plant was designed for two turbogenerators, each of 400-kW capacity, but only one was installed. Its purpose is to provide information on problems that would be encountered with the large developments required to harness the enormous tidal potential along the White Sea coast of Russia.

A rockfill dam across Jiangxia Inlet, about 200 km south of Hangzhou on the east coast of China, had been built for reclamation purposes and aquaculture but was altered in 1980 to incorporate a tidal power plant. It has an installation of 3.9 MW provided by five bulb-type and one straight-flow (STRAFLO) type turbogenerators.

A 20-MW tidal-electric plant was commissioned in August 1984 at a site in the Annapolis Basin of Nova Scotia, Canada. The purpose of this installation was to evaluate the operational characteristics of a large-diameter (7.6-m) STRAFLO turbine and the use of such turbines for the large tidal energy potential at sites in the Bay of Fundy, Canada, as well as for low-head river developments. The output of the plant feeds directly into the transmission system of Nova Scotia Power Inc., the provincial utility, and its subsequent operation has been successful.

A brief description of these existing developments is presented in Chapter 15.

A major shift in attitudes to energy developments has also provided a substantial impetus to the reappraisal of this renewable energy source for use in large, modern, electrical utility systems. The technology is now available and proven for exploiting the energy of the tides. There are no fuel costs since a tidal power development uses a freely available, renewable source of energy so that tidal energy is virtually inflation free, except for operating and maintenance costs. The tides are an energy source completely and accurately predictable as far into the future as it is necessary to consider and, of particluar significance, the harnessing of the tides for electrical energy generation is nonpolluting!

Electrical energy plays a very prominent role in the growth process of any economy, whether it is developing or developed. It is needed for industry, agriculture, transport, domestic, and many other activities that create employment, produce goods and services, and generally affect the day-to-day lives and the standard of living of the people. The grave concerns for the environment that began to surface in the 1970s, and that have since been given increasing attention by nations, stems largely from the fact that the world has come to depend predominantly on burning hydrocarbons to supply energy needs. Moreover, the uncertainties associated with "conventional" energy sources have highlighted the necessity for diversification and expansion of the energy resource base.

The dramatic increase in energy prices of the 1970s brought about a sharp expansion of the energy resource base, while the steady increase in the carbon dioxide content of the atmosphere has made it imperative to seek ways and means to limit the production of fossil fuels. These are the major contributors to carbon dioxide build-up, which is now resulting in a global increase in atmospheric temperature and other environmental threats. As a result tidal-electric developments are now potentially economical, viable energy sources for consideration in the expansion plans of those electrical utility systems, fortunately situated along coastal regions with high tides and appropriate coastal configurations, and faced with the effects of population and industrial growth.

As with any development by society the exploitation of tidal energy will have some ecological effects that will be minimal compared with the impacts of population growth resulting in urban sprawl, proliferation of transportation routes, increased energy demand and the like. The barrage and hydraulic control works will perform satisfactorily for a century or more without large capital replacement expenditure and, with timely and effective maintenance, their usefulness can be extended indefinitely. This compares with the useful life of 30 to 40 years for nuclear and thermal power facilities. Moreover, a tidal-electric development has low operation and maintenance costs so that the production of its energy is relatively inflation proof. Therefore, within the energy field, a tidal-electric development offers **sustainability** with a long useful life, no fuel cost, no environmental pollution loading, and reduced energy costs over the longer term.

CHAPTER 2

TIDAL PHENOMENON

2.1 INTRODUCTION

On the shores of an ocean, the water level rises and falls—a fluctuation that occurs twice each day, that is, semidiurnal, or more precisely, twice each 24 h, 50 min, and 28 s—the apparent period of rotation of the moon. Primarily due to the upward gravitational pull of the moon, the water level will gradually rise until it reaches its highest point, termed *high water* and then fall untill it reaches its lowest point, or *low water*, when it turns and rises again. The difference between these two points is called the *range* of the tide.

Also, where there is a long bay, an estuary or a river mouth on the coast, the water, when high, will flow inland into it and, when low, it will flow out again. The inflow during the rise of the tide is known as the *flood*, and the outflow during the fall of the tide is known as the *ebb*. Slack water occurs at the top of the tide and at the bottom. Thus, there are two distinct movements that are ordinarily included by the word *tide*: A vertical rise and fall in the level of the water that is the *tide* proper, and its horizontal flow in the two directions, ebb and flow, alternately, distinguished by the term *tidal stream*. It is important to distinguish clearly between these two distinct aspects. The direction and velocity of the tidal stream at the site of a proposed power development will have a decided influence on the design and construction of all permanent works, as well as on the method and stages of construction.

As defined, the range of the tide is the difference in level between successive high and low waters. In any tidal power scheme it is the range of the

Elements of Tidal-Electric Engineering. By Robert H. Clark

tide that determines the potential head available for powering the turbines, although the sluiceway capacity and operating characteristics of the development will have a bearing on the amount of this head that is effectively used.

The intimate relationship of the tides with the periodic movement of the moon is well known, but the tidal phenomenon observed at different locations is complex. Therefore, since the laws of the tidal cycle are of great importance for the solution of the problem of tidal energy exploitation and utilization, the basic concepts and terminology of tidal theory are discussed briefly here.

The amount of rise and fall of the tide is not the same everywhere; it varies from almost nothing to a range of over 17 m in some locations. On the other hand, in all parts of the world, the time of high water is not at the same hour of the day but, on the average, it occurs about one hour later from one day to the next and in a period of 14 to 15 days it comes around to the same hour again. At every locality in every ocean there is also to be found a well-marked variation in the range of the tide, which usually recurs twice in the course of each month. This variation may be very different in its character from region to region; for example, it may be a change in the range of the tide from a large difference in level to a small difference, or the two tides of the day are sometimes exactly equal in their range and at another time one of the two is much greater than the other, alternatively. There are some localities where at certain times of the month the tide becomes *diurnal*, that is, with only one high water and one low water during the day, but the reason for such exceptions does not conflict with the general principles. However, whatever the character of the variation may be, the change always recurs in a period of either about 2 weeks or about a month in every ocean.

2.2 OCEAN TIDES

The tide in the various oceans progresses through them as an undulation; for example, the tide runs up the Atlantic Ocean as an undulation from its southerly end between South America and Africa, to its northerly end between Canada and Europe (Dawson, 1920). The range of these oceanic tides is usually not more than about 1 m.

The ocean tide is produced primarily by fluctuations in the resultant of the several forces of gravity and of centrifrugal action caused by Earth's rotation and by changes in the relative positions of earth, the moon, and the sun. Thus, the force causing the tide at any point on Earth is substantially the resultant of the attractive forces due to the moon and sun and the centrifugal force caused by rotation of the Earth–moon system (Dean, 1966). Because it is so much nearer Earth than the sun, the moon exerts a much greater influence on the tides than does the sun, despite its comparatively small size, since attractive force is proportional to the inverse square of the distances between the bodies. In theory, the solar tide is about 45 percent of the lunar, varying from about 38 percent at perigee to about 52 percent at apogee (refer to Fig. 2.3). Hence, a knowledge of the relative movements of the sun, moon, and Earth is necessary for an understanding of the tides, and particularly of the variations in their range that take place in a systematic manner over a long period of time.

2.3 TYPES OF TIDES

Every movement of the sun and moon has its effects on the tide but, in different regions of the world, a particular movement of the moon may have a dominating effect and others become of secondary importance. Moreover, the relative influence of the sun and moon are different in different parts of the world. Taking into account these variations, all tides may be grouped into three principal types, according to the dominant feature they represent. These three types are *synodic, anomalistic*, and *declinational* tides. Although one of the variations may be dominant, the others are never entirely absent and may be of sufficient magnitude to result in a *mixed* tide.

2.3.1 Synodic Tide

The synodic tide occurs when the dominant variation in the range of the tide takes place twice a month; the range being greater at new and full moon and less at the moon's quarters.

When Earth, sun, and moon are in the same straight line, which happens about twice a month, the attractions of the moon and sun act together, and the range of the tide is much greater than when the moon, Earth, and sun are in quadrature. Thus, the time of the new moon and full moon produces *spring* (or high) tides, and when the moon is at its quarters, *neap* (or low) tides occur. These relationships are illustrated in Figs. 2.1 and 2.2. The time from new moon to new moon is known as the *synodic* month. Because of the moon's perturbations, that is, the irregularities in its movement due to the sun's attraction, the exact length of the *synodic month* varies, but its average value is 29.53 days.

2.3.2 Anomalistic Tide

The anomalistic tide results when the greatest variation in range accords with the moon's distance and takes place once a month. (See Table 2.1.)

The path of the moon as it revolves around Earth is not circular but elliptical. Moreover, Earth is not in the center of the ellipse, as might be supposed, but is off center. Therefore, during each revolution of the moon around Earth, there is one point, and one only, where the moon is nearest Earth. This is known as *perigee* (Fig. 2.3). The point where the moon is farthest from Earth is called *apogee*. The period of time from perigee to perigee is the *anomalistic month*, which has an average length of 27.55 days.

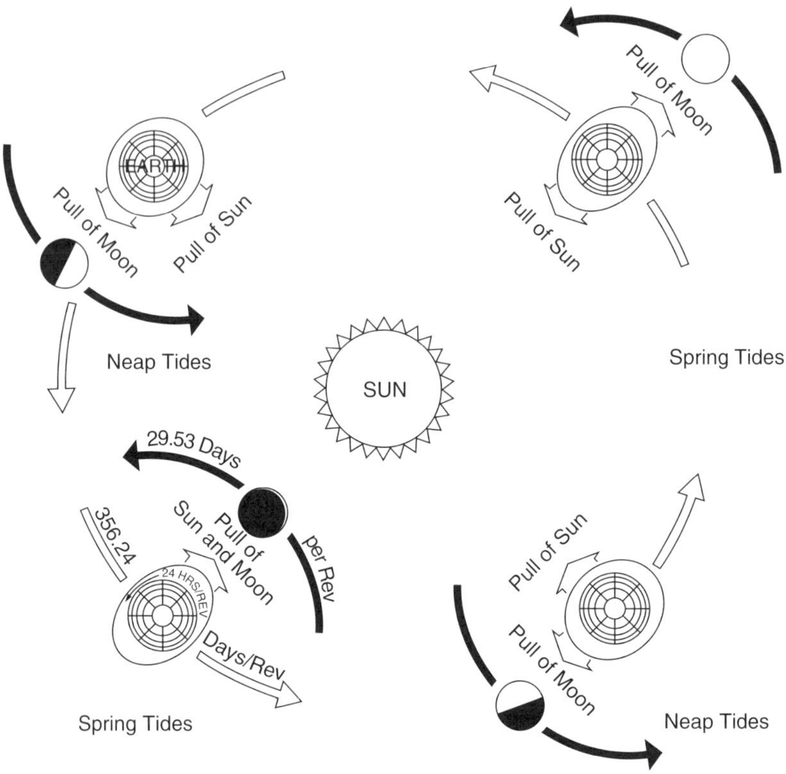

Figure 2.1 Origin of the tides.

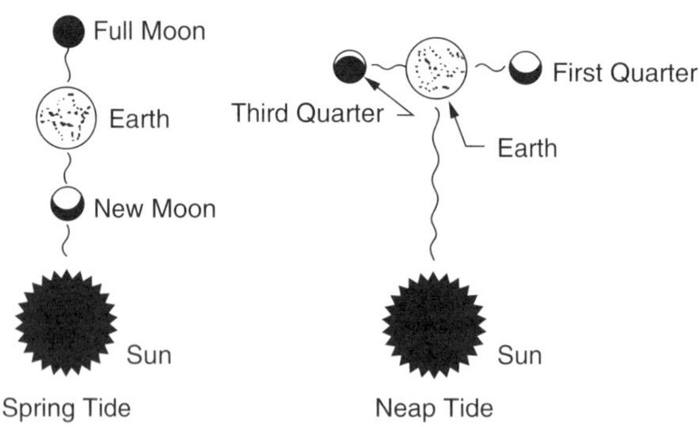

Figure 2.2 Conditions for spring and neap tides.

Although there do not appear to be any tides that are as purely of the anomalistic type as those of the synodic type, there are regions where this variation with the moon's distance is distinctly greater than the variation from springs to neaps. This is the case in the Bay of Fundy as shown by the following table giving the variations at Saint John, New Brunswick, about half way

TABLE 2.1 Comparison of Anomalistic and Synodic Effects

	Saint John, N.B.		Burntcoat Head	
Description of Tide	Range (m)	Difference	Range (m)	Difference
At perigee: range at spring tides	8.11		15.39	
At apogee: range at spring tides	6.07	2.04	12.25	3.14
Mean spring range	7.09		13.82	
Neap range at moon's mean distance	5.50	1.59	11.82	2.00
Average range during mean	6.29		12.82	

Source: From Dawson (1920).

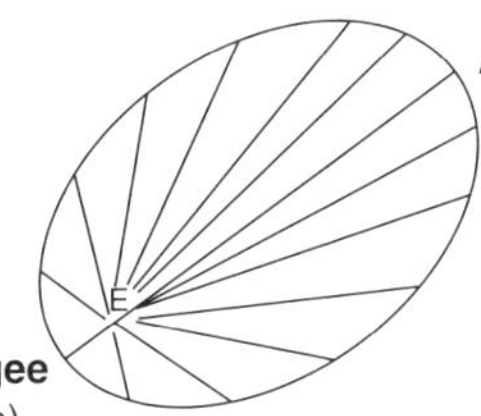

Figure 2.3 Anomalistic tidal variation.

along the bay, and at Burncoat Head in the upper reaches of the bay (refer Fig. 16.7).

2.3.3 Declinational Tide

This type of tide results when the changes due to the moon's declination (which makes the two tides of the day unequal in range) are so large and obvious that all other features of the tide are obscured.

The position of the sun at its zenith varies from day to day. In the northern hemisphere, it is higher in the summer than in the winter since the plane of the equator does not coincide with the plane of Earth's orbit. This angular position of the sun with respect to Earth's equator is called its declination. In a precisely similar way, the position of the moon varies, resulting in the effects illustrated in Fig. 2.4. The period of time that elapses between the successive times that the sun crosses the equator in the same direction is, of course, about one year, whereas for the moon, it is about one month or, more precisely, 27.32 days. This period is known as the *tropical month*.

In regions where the tides are dominated by the moon's declination, the variations that result are greater than from any other movement of the moon considered separately. When the sun and moon are on the equator, the two tides of the day will be exactly equal in height. When they are not at the

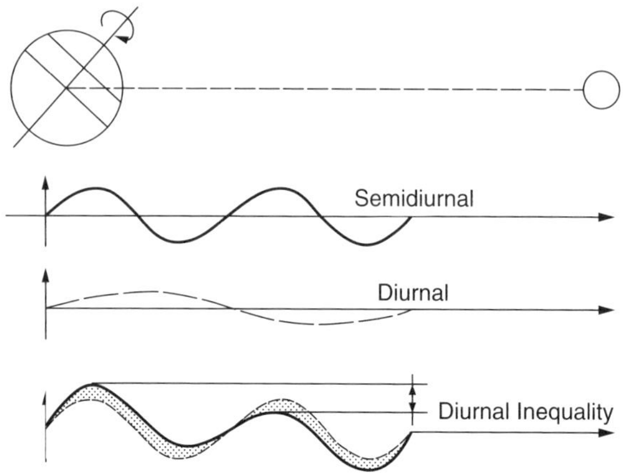

Figure 2.4 Effect of moon's declination.

appropriate declinations to balance their respective declinational influences, there is a *diurnal inequality*.

2.4 PROPAGATION OF TIDES IN ESTUARIES

An estuary is small enough so as not to be subject to measurable tides on its own account. When the tide leaves the open ocean and enters straits, inlets, bays and, estuaries, it undergoes a change that is more closely related to the laws of hydraulics and wave motion than to its astronomical aspects. An elaboration of the physical understanding of tidal flows in estuaries is presented by Ippen and Harleman (1966).

Throughout the ocean, the tide is almost always a symmetrical undulation, having the same form as the long swell of the ocean; the summit of the undulation causing high water and the trough causing low water. On entering an estuary, the form of the undulation undergoes the same kind of modifications as waves do in shallower water near the shore. The observed sea level is raised and, depending upon the configuration of these inlets, the amplitude of the tidal wave may be increased many times (the term *tidal wave* should not be confused with the *tsunami*, a seismic phenomenon). Such a wave tends to move along an estuary with a speed given by $\sqrt{gd}$, where d = mean depth. To illustrate this, if the mean depth of an estuary were 50 m, then the wave length L = wave period × speed = $44{,}700 \times 22.14 \times 10^{-3}$ = 900 km (approx.). The period of a semidiurnal ocean tide of 12 h and 25 min (12.42 h) is long enough to make the wave length much longer than the physical length of an estuary. Although most estuaries are shorter, one with a length that is about one-quarter the wave length would invoke a resonance response so that the amplitude of the tide could increase along the estuary by factors of four or more.

For example, the overall resonant frequency of the Bay of Fundy is essential to the production of the very high tidal amplitudes at the head of the bay (refer to Fig. 7.3). The narrowing and shelving of the Bay of Fundy brings the whole system from the Continental Shelf through the Gulf of Maine to the upper reaches of the Bay of Fundy to near resonance with the semidiurnal tides. From a comparison of the principal solar and lunar ecliptic tidal harmonics of the Bay of Fundy/Gulf of Maine system with amplitudes in the North Atlantic Ocean, it has been determined that the system behaves as if its natural period were approximately 13 h. Thus, along the outer edges of George's Bank, the oceanic tide has a mean range of about 1 m; at the mouth of the Bay of Fundy, the tidal range has increased to about 3.5 m (at Yarmouth, Nova Scotia), to a mean range of about 13 m in Cobequid Bay and 10 m in Cumberland Basin (refer to Fig. 7.3). The tide at Cape Maringouin occurs about 40 min later than at the mouth of the bay; within Minas Basin, the time of high water occurs even 40 min later than it does outside that basin.

2.5 CORIOLIS EFFECT

A body moving over Earth is deflected from a straight path relative to Earth's surface because of Earth's rotation. In the northern hemisphere, the force causing this deflection, known as the Coriolis force, is always to the right. Thus, for example, the water in the Bay of Fundy tends to pile up earlier along the Nova Scotia shoreline on the flood tide and thus has an influence on the tidal range at a particular location.

The Coriolis force results from the conservation of angular momentum. As a fluid stream moves north or south from the equator on the rotating Earth, it moves nearer to Earth's axis and thus has an angular momentum greater than that required. To conserve its angular momentum, its angular velocity is increased, causing the stream to be deflected eastward until friction and hydrostatic forces stabilize it. Similarly, as the stream moves toward the equator, it moves away from the axis, requiring additional momentum so it swings or slopes to the west. The term, $2\omega \sin \phi$ is known as the Coriolis parameter, where ω is the angular velocity of Earth's rotation about its axis (7.27×10^{-5} rad/s) and ϕ is the latitude.

2.6 BARRAGE EFFECTS

The most important effect, from the point of view of tidal-electric engineering, is that which a barrage or tidal development may have on the resonance of an estuary. The barrage could increase the tides at the barrier because it would bring the foreshortened estuary closer to resonance, or it could reduce materially the tidal range by upsetting the natural resonant condition. Thus, it is very important early in an investigation to be able to simulate the natural tidal regime and then to determine the effect on this regime by the placement of a barrage at various locations within the estuary as well as by the operation of the development.

The mathematical modeling of the tidal regime to determine the effects of the construction and operation of a tidal-electric plant on that regime is discussed in more detail in Chapter 7.

CHAPTER 3

TIDAL POWER POTENTIAL AND SITE SELECTION

3.1 HYDROELECTRIC VERSUS TIDAL-ELECTRIC DEVELOPMENTS

River hydropower and tidal power may have many characteristics in common, but there is a fundamental difference in the forces activating these two energy sources. These forces are the downward-acting terrestrial gravitation causing the water to flow on Earth's surface, and the upward-acting lunar gravitation causing the water to oscillate in estuaries. There has been a tendency to ignore this basic dissimilarity in the driving forces and to treat tidal-electric developments simply as low-head hydroelectric developments with added complications.

The cardinal feature of the tides is that, regardless of their local variation, they occur in a most orderly fashion and are predictable. Both the periodicity and predictability of tidal action are important characteristics bearing strongly on the utilization of this phenomenon as an energy source. Tides at any location repeat themselves almost identically about every 19 years. Within any 1-year interval, the differences are relatively small so that the available energy is practically the same from year to year. Moreover, at least for the semidiurnal tides, there is a basic relationship between the tidal range and hours of high and low tide so that, at a particular location and at a specific time during the day, the tidal range to be expected will always be within known values. To illustrate, the upper and lower curves of Fig. 3.1 envelope the plots of all tidal range values versus the time of occurrence of high tide at Economy Point in Minas Basin, Canada (refer Fig. 16.8). This relationship shows that, at this

Elements of Tidal-Electric Engineering. By Robert H. Clark

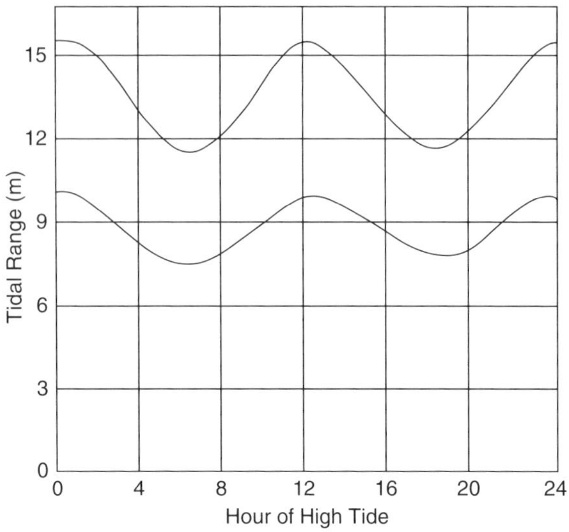

Figure 3.1 Relationship between tidal range and hour of high tide in Minas Basin.

location, the tidal range to be expected when the hour of high tide is at 0600 hours will never be greater than 11.5 m nor less than 7.6 m.

The water quantity involved in a tidal cycle is relatively constant so that, with tidal energy, there is no wet or dry year and, even more important, no wet or dry month—no summer or winter. In contrast with the foregoing characteristics, river energy production is fraught with unpredictable wet and dry cycles. Moreover, it is possible to predict the amount of tidal energy available in a week, a month, or a year, except for minor variations due to meteorological influences on the water surface.

The approaches to the hydraulics and to the machinery and plant design and construction are also different for river hydro and tidal power developments. The most economical design for the former development is determined by the hydrological characteristics of the river such as its minimum recorded flow, flood flows, storage possibilities, and head. For a tidal development, on the other hand, the design can be varied to produce the desired capacity, which would be limited by economic considerations rather than by available flow. Thus, the output of a tidal plant is determined by the following interrelated factors: (1) usable head that varies continuously with the tidal regime and is modified by the fluctuation of basin levels resulting from operation of the plant; (2) the area of the tidal basin; (3) the capacity of the sluices used to fill or empty the basin; (4) the capacity of the generating units; and (5) the method of operation selected. Because of the abundance of water available to a tidal plant, emphasis of design is more on maximum output than on maximum efficiency, the latter being of more importance in river hydro developments. Thus, the conversion of tidal energy differs completely from that of river hydroenergy where it is usually not too difficult to ascertain the order of magnitude of the head, mean discharge, and spillway capacity. Because of its continually changing head, it is impossible to determine a priori the dimensions of the equipment and the magnitude of the installed capacity for a tidal-electric development. Optimization is a fundamental requirement in the design of a tidal power project.

The spillway gates of a river hydroplant may have to be operated only a few times annually, whereas for a single-basin, ebb-generation tidal power scheme, the sluiceway gates would be operated at least 1400 times annually and about twice that for double-effect operation. The sluiceway capacity of a tidal plant is selected to allow, alternately, the filling and emptying of the basin to minimize the unit cost of energy rather than to act as a relief valve to prevent overtopping of the dam, as in the case of a river hydroplant.

3.2 SITE POTENTIAL ESTIMATION

The determination of the appropriate magnitude of installation in a tidal power plant is a good deal more complex than for hydroelectric undertakings. Duration curves of flow and head yield a fairly accurate estimate of the average annual energy available at a potential river power site. Available pondage, peak demand, the character of the load–duration curve, and the kinds of generation already available in the system determine plant-installed capacity. For a tidal site, the available energy (and peak capability) depends largely on the installed capacity within very wide practical limits. Basin size and tidal range do, of course, define the theoretical or gross energy potential (GEP) of a site. The formula for computing this value is derived, with reference to Fig. 3.2, as follows:

For a tidal basin at some height z above low tide

Energy to fill is:
$$E_f = \gamma g \sum_{z=0}^{z=Rz} zA_z\,dz \tag{3.1}$$

where A_z is the area at height z above datum (minimum tidal level), γ is the density of seawater, and other symbols are as defined in Fig. 3.2.

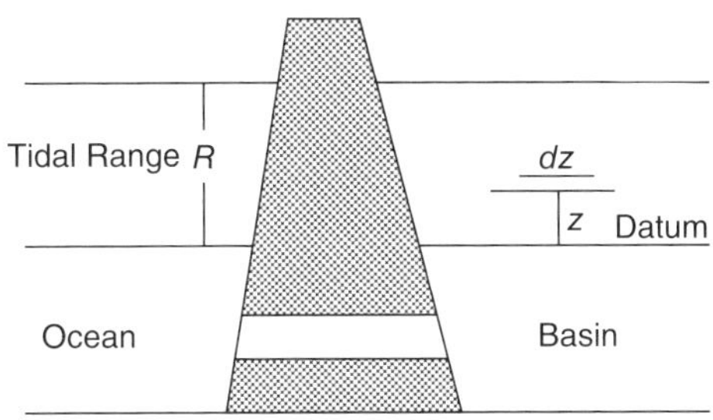

Figure 3.2 Derivation of power formula.

For the particular case where A_z is a constant and independent of z,

$$E_f = A\gamma g \sum_{z=0}^{z=R} z dz = \frac{\gamma g A R^2}{2} = \gamma g \frac{VR}{2} \tag{3.2}$$

where V is the volume of the basin.

Therefore, the energy for the complete cycle

$$E = \gamma g V R \tag{3.3}$$

If A is the area of the basin in square kilometers and R the tidal range in meters, then the Gross Energy Potential (GEP) is

$$\begin{aligned} \text{GEP} &= 1.025 \times 981 (A \times 10^{10})(R \times 10^2)^2 \times 10^{-7} \text{ W-s} \\ &= 1.005 \times 10^{10} A R^2 \text{ W-s} \\ &= 2793 A R^2 \text{ kWh per tide} \end{aligned} \tag{3.4}$$

For 705 tides per year,

$$\text{GEP} = 1.97 \times 10^6 A R^2 \text{ kWh annually} \tag{3.5}$$

$$\text{Average power over tidal cycle} = 225 A R^2 \text{kW} \tag{3.6}$$

It is significant that tidal energy varies with the square of the tidal range. Moreover, it is evident that the theoretical average operating head is R/2, or the amplitude of the tide. Obviously, this value cannot be realized in practice over the whole cycle, although it can be approached for limited periods under single-basin, ebb-tide operation (refer Chapter 5).

Although Eqs. (3.2) and (3.3) are independent of basin shape and tide symmetry, the underlying derivation of Eqs. (3.5) and (3.6) is significant to their proper application. First, the basin area during the tidal cycle was assumed to be constant; second, the equations would apply to basins whose dimension in the direction of propagation of the tidal wave is equal to or smaller than the tidal wave length; and third, the possible influence of the form of the basin bottom, and its overall shape, on tide propagation is not accounted for. However, if the basins are sufficiently deep and with fairly high tides, then Eqs. (3.5) and (3.6) will provide an order-of-magnitude estimate of the tidal energy potential at a site and may be useful in a preliminary parametric comparison of various sites.

Tidal engineering practice, based on the results of preliminary studies of tidal sites at various locations on the globe, usually allows an optimal annual energy production (AEP) for a single-basin site generating on the ebb flow of about one-third of the GEP. It is emphasized that the value obtained is an order-of-magnitude estimate only.

3.3 COEFFICIENT OF THE TIDE

For power development, the highest tidal ranges are required and the tides should be nearly uniform in both their high and low levels in successive cycles. Synodic and anomalistic tides are preferred, particularly the latter. Declinational tides are not as satisfactory.

As explained in Chapter 2, the highest synodic tides occur when the moon and sun are in conjunction or opposition, the highest anomalistic tides take place when the moon is in perigee and the maximum declinational tides are realized with the moon nearest to the zenith or nadir. Successive synodic tides vary to some extent in height, becoming uniform only near the times of the moon in its "quarters" in contrast to the much greater uniformity between successive anomalistic tides at all times. For example, the synodic tides on the Brittany coast of France (Fig. 16.13) and the anomalistic tides in the Bay of Fundy, Canada (Fig. 16.8) are considered to be suitable for power development.

Successive declinational tides, however, exhibit a very large variation in amplitude. For example, when the moon is in highest declination, there is only one tide per day—either a high tide or a low tide, following which the water remains at the opposite level (low or high, respectively) for about 8 h exhibiting only relatively small oscillations. The extreme effect occurs when both moon and sun are at their highest declinations at the same time. These large diurnal inequalities tend to make the declinational tide less satisfactory for energy exploitation.

It is interesting that these three types of tide can occur within a relatively small region (Dalton, 1961). For example, synodic tides are in evidence along the Atlantic coast of Nova Scotia, anomalistic tides in the Bay of Fundy and Minas Basin, and declinational tides in the Northumberland Strait. These locations are shown in Fig. 16.8.

In calculating tidal power available at any site, the term *coefficient of the tide* may be used to access the relative range (Gibrat, 1966). Typical coefficients are:

Exceptional equinoctial high tides	1.20
Average equinoctial high tides	1.00
Average high tides	0.95
Average tides	0.70
Average low tides	0.45
Feeble low tides	0.20

Under equinoctial conditions, on which these coefficients are based, the sun is at its average distance from Earth and at zero declination.

3.4 MAJOR FACTORS INFLUENCING PROJECT ECONOMICS

The important factors to be taken into account in the prefeasibility investigation may be broadly classified as *external* and *internal* factors. The major factors are listed in Table 3.1. Some of the external factors may be quite general, such as the future prices of fuel, and would apply to a consideration of any generation source to be added to a utility system. Some external factors are site specific, such as tidal range, and may be defined quite accurately by appropriate surveys or measurements while the value of others are the results of judgments, such as the main thrust of a future expansion of the utility system, and their accuracy will not be known for several decades.

Many of the internal factors listed are interdependent and may also depend on some of the external ones; for example, bathymetry, geology, and availability of suitable dredging plant affect directly such internal factors as: number and size of turbines, turbine caisson foundation elevations; turbine submergence, method of turbine operation, number and size of sluices, and length and alignment of barrage. These, in turn, affect the tidal range on the seaward side of the barrage, the energy output of the plant, cost of development, and so forth, so that the final design for a particular scheme would be the result of successive refinements of the values of the foregoing factors.

3.5 SITE SELECTION

3.5.1 Site Identification

The identification of potential sites is a first step in the reconnaissance of those estuaries that have sufficiently high tides to offer possibilities for tidal power exploitation. This step should make use of available maps and charts showing bathymetry, topography and geology of the region; data on tidal ranges along the estuary, which may be deduced from various sources and physical evidence; and any other information relevant to an energy development in the region. Numerous other factors may also influence site identification, such as the desirability (or requirement) of including sites that might have only a small potential energy output or that could have a multipurpose aspect.

3.5.2 Preliminary Assessment

This step in site selection would only be for the purpose of evaluation of the potential for development and to identify the most promising site locations. Therefore, for this purpose and at this stage in site selection, it should not be necessary to identify sites that would appear to be very similar in location and size and with similar potential development problems.

Bathymetric data are necessary in order to provide, at the minimum, approximate estimates of the active volume of the tidal basin, the size of closure dam, and the depth of powerhouse foundations that would be

TABLE 3.1 Major Factors Influencing Project Economics

External Factors	Internal Factors	Results Affected	Conclusions Regarding
Future prices of fuel	Turbines: type, size, number, and cost		
Development of electrical utility systems	Sluices: type, size, number, and cost		
	Generator capacity and cost		
Tidal range Wave climate	Turbine submergence and cavitation criteria	Energy output	
Bathymetry	Method of operating turbines		
Availability of suitable dredging plant	Turbine caisson foundation elevations		**Technical feasibility**
	Turbine draft tube geometry		
Distribution of existing erodible sediments	Barrage length and alignment		
Supply of fresh sediments from outside of estuary	Effect of barrage on tidal range	Construction costs	**Economic and financial performance**
	Design criteria for structures, including risks		
Availability and costs of construction materials	Rate of turbine production		
Availability and cost of capital	Construction sequence and closure velocities		**Acceptability**
Availability, productivity, and cost of labor	Direction and strength of current velocities from turbines and sluices	Environmental and social aspects	
Present and future water commerce	Caisson construction yards—location and cost		
	Design of embankments		
	Design of transmission links		
Differential inflation (e.g., between energy prices and construction costs)	Navigation locks—location, size, and cost		
	Effects of barrage on ship movements		

Source: Adapted from: Severn Barrage Committee (1981).

necessary relative to the available tidal storage behind the dam. A plot of available tidal range information along the estuary would yield information on tidal ranges at identified site locations where specific tidal data are not available. At this stage, geotechnical and environmental constraints at a location could probably only be given general consideration, but locations where abutment conditions are known to be obviously poor should, for example, be avoided.

Figure 3.3 identifies the major steps in the process of site selection for preliminary assessment. With reference to the figure, site-specific data would include consideration of geotechnical conditions, structure lengths, access, transmission line routings, possible environmental factors, and any navigation requirements. Based on a determination of the order of magnitude of the probable optimal annual energy production, the installed capacity could be approximated by using a tentative capacity factor of one-third, which is about the value obtained from the Fundy (Canada) and Severn (United Kingdom) prefeasibility studies for ebb generation. If, at a specific location, there is a significant diurnal inequality, then the capacity factor should be reduced appropriately; in other words, a greater installation would be required to optimize the output. The capacity of the installation would be given by

$$\text{Capacity (MW)} = \frac{\text{kWh} \times 10^6}{\text{capacity factor} \times 8760 \times 10^3} \tag{3.7}$$

Parametric analysis could be used at this stage to avoid detailed cost estimates and layouts at each identified potential site in order to compare and

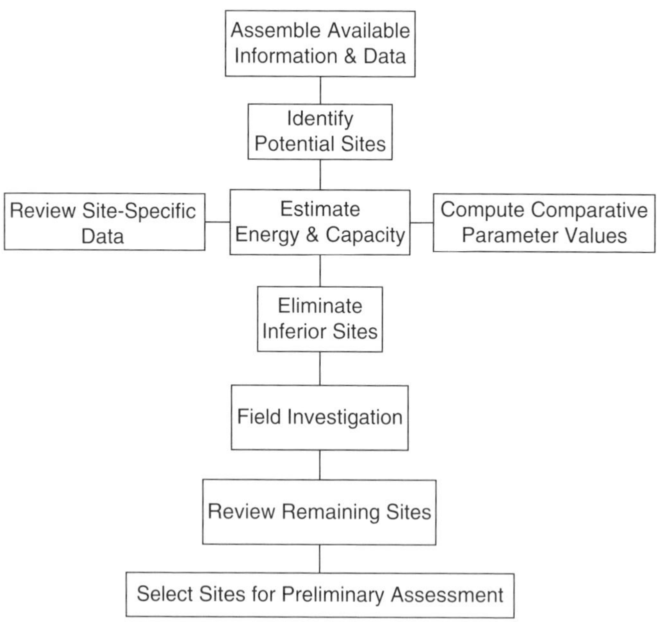

Figure 3.3 Site selection for preliminary assessment.

eliminate many sites from further consideration. Such an analysis would likely suggest the probable order in which potential sites should be costranked. Obviously, it would not be possible to produce reasonably firm cost estimates at any particular site until site-specific configurations were developed, which would obviously depend upon reasonably adequate bathymetric and geophysical data.

For such a parametric analysis, for example, the relative capital cost of a development may be represented by the product of the total length, L, of the development and the square of the height, H, of the structure at the deepest point, that is, $L \times H^2$. If this product were divided into the probable, optimal annual energy production in kilowatt-hours, the result would provide a rough basis of comparison of the at-site cost of energy for the various sites initially identified. Higher values of this parameter would indicate generally favorable economic benefits. Obviously, this parameter is based on the generalized assumptions that the capital cost of a tidal-electric plant is proportional to the cost of civil works and that the civil costs are roughly proportional to the volume of the barrage. The cost of turbines and other mechanical and electrical equipment could be assumed to be constant for comparable tidal ranges and plant outputs. However, for lower ranges, the cost per megawatt of the electromechanical equipment would probably be appreciably more than for higher tidal ranges. Other parameters may be more appropriate for the preliminary site elimination process or the number of possible sites may be so few in number that this stage may be bypassed.

The information and value judgments on various factors may be set out in a *conflict diagram* to indicate in a qualitative manner, by visual analysis, the relative merits of each site. Such a diagram is shown in Fig. 3.4 for several obvious factors; others could be added to reflect specific regional conditions.

One obvious factor that could be added to the diagram would be *barrage effect* on tidal regime. It may be fairly obvious from the physiography of the

Factor	Site										
	A	B	C	D	E	F	G	H	I	J	K
Relative Cost	■	■	■	■	■	■	◪	◪	□	□	□
Closure Criteria	■	□	□	■	■	■	◪	■	□	□	□
Causeway Potential	■	■	■	■	■	■	□	□	◪	◪	◪
Access	◪	□	□	■	■	■	□	◪	□	□	□
Environmental Issues	◪	■	■	◪	◪	■	◪	◪	◪	◪	□
Probable Foundations	◪	□	■	◪	□	◪	◪	□	□	□	◪
Navigation Issues	◪	◪	◪	◪	□	◪	□	■	□	□	□
Transmission Tie	◪	□	■	◪	◪	◪	□	□	□	□	□
Seismicity	■	◪	◪	■	■	■	□	◪	◪	◪	◪

Legend: ■ Major constraints ◪ Minor constraints □ Favorable conditions

Figure 3.4 Summary of selection criteria evaluations.

estuary, the location of the sites, and the tidal data available that the barrage sites identified would not have a major impact on tidal levels. However, the sensitivity of energy generation at the identified sites to changes in tidal range can be assessed approximately by very preliminary calculations based on formulas described by Ippen and Harleman (1966). This rather simplistic calculation would yield possible changes in tide levels due to barriers at the identified sites.

In lieu of basing the parametric ranking on the order of magnitude estimate of annual energy production from Eq. (3.5), a relatively simple, and computationally fast, mathematical model is available to provide a fairly close estimate of annual energy production. This model is described in Chapter 10.

3.5.3 Short Listing of Sites

From the initial sites identified and subjected to parametric cost ranking and other appropriate considerations, the results of the comparative evaluations may identify several sites as meriting further consideration. For these, specific configurations should be developed, taking into account, more precisely, foundation conditions for which field survey programs may have to be carried out to provide adequate information. However, at this stage, it would be necessary to collect only such data as are necessary to be able to make a reasonably firm cost estimate, but accurate enough so that a decision on its results could not be overturned on technical and scientific grounds.

Generally, the budget and time for such preliminary assessments and prefeasibility investigations are limited so that it is not possible to evaluate and optimize each scheme fully. Nevertheless, some of the more important variables, such as those listed in Table 3.1, should be taken into account. Obviously, this is a further stage in the site selection process and should be designed to provide a more refined ranking. Again assumptions will have to be made regarding sizes of turbines and sluiceways. The likely unit costs for principal construction materials and elements should be assembled, including turbogenerators, reinforced concrete, and wave armoring (the latter might be based on marine costs elsewhere in the region). The following steps are suggested to provide reasonable cost estimates and energy production figures in order to arrive at a shorter list of schemes: (1) prepare outline designs for main elements of the barrage; (2) assess energy output and, if appropriate, input energy for pumping on an annual basis for a range of turbine and sluiceway sets that would include the optimum set in terms of cost of energy; (3) assess likely value of energy output taking into account alternative system plant mixes; (4) assess difficulty of closure; (5) prepare likely construction schedules for each scheme in order to assess the effect on the costs; (6) calculate benefits and costs; and (7) assess risks and uncertainties.

As a result of the foregoing activity, a very short list of schemes, three at the most, would be finally selected for refined studies and selection of "best" site and its configuration for the estuary and electric utility system.

CHAPTER 4

MANAGEMENT AND ORGANIZATION OF INVESTIGATIONS

The organization of a composite management structure that brings to bear the many interests and authorities involved in feasibility and planning studies of tidal power development is not a simple task. The problems of setting up the appropriate organization for the management of studies compound the already difficult task of evaluating the viability of developing a tidal power site that is a low-head development and where uncertainty exists and capital costs are inherently high. As illustrated by Fig. 4.1, such studies involve a complex interrelationship of many activities. The complexity is even further compounded if economic viability relies on multipurpose development that introduces the not insubstantial infrastructure of urban industrial and transportation facilities. And finally, there are the vitally important environmental concerns.

4.1 MANAGEMENT

At some tidal power sites in the world, economic considerations may involve broader regional issues than power itself. Harbors, ports, industrial development, transportation, communications, water supply, agriculture, fisheries, scientific development, and other interests may play significant roles in justification of a development. Nevertheless, the leadership role and focal point for national concentration on the prospective development is clearly the

Elements of Tidal-Electric Engineering. By Robert H. Clark

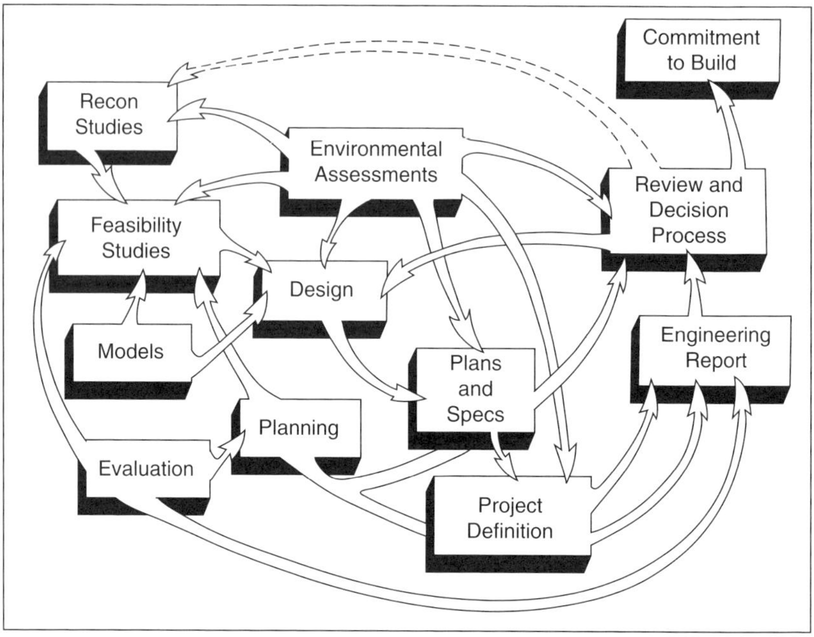

Figure 4.1 Complexity of tidal power studies. (From Warnock, 1978.)

responsibility of that sector of the economy most directly involved in the prime justification for the project, namely the energy and power sector with which the identifiable revenue and financiability rests. Figure 4.2 indicates some of the inputs by various sectors of the economy that a multipurpose tidal power development may involve.

The interests involved in a tidal power investigation might be broadly classed as the *sponsor*, *other participants*, and *support groups*, linked as shown in Fig. 4.3, for an effective organization and management. The sponsor's role is primarily to provide leadership and should rest with the sector having the primary revenue earning interest.

The steering group, or sponsor's team, would have the primary responsibility for the identification of all risks, hazards, and opportunities in the creation of a viable profitable venture. This group would thus have the obligation to advise abandonment of a project should any reason emerge to prejudice the true worth of the undertaking in the long-term context.

It is at the steering group level that the critically involved interests, such as ocean trade, industry, regional development, fisheries, and environment would play their part in the decision process. Their representatives should be of a caliber and distinction to present judgments that will be respected and upheld. Finance and science and technology, vital to a national project, would be firmly linked to the action groups.

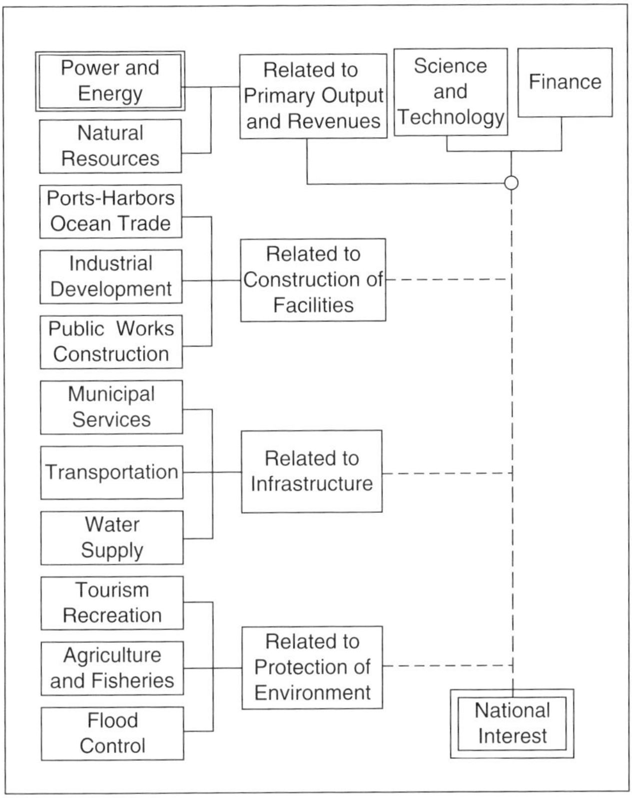

Figure 4.2 Inputs to a multipurpose tidal power development. (From Warnock, 1978.)

The study must have top experts working on specific aspects of the project, but if the organizational and management framework does not lend itself to the molding of an effective team, then much of the effort will be nonproductive and the expertise lost. Moreover, it should be assured that every significant area of expertise needed for a particular project is adequately covered by experts entirely familiar with, and dedicated to, the multidisciplined approach. A tidal power project requires, from beginning to end, a continual synthesis of previous thinking and achieved results that only a person, disengaged from partial studies but competent enough in every domain, can do.

On major giant projects, such as the Churchill Falls Hydroelectric Power Project, Canada (Warnock, 1979) or a tidal-electric power project, the tasks assigned to the steering group make it essential that it remain separate and independently identifiable from project management. The latter activity may be "contracted out" and involves the tasks of implementing the sponsor's plans and completing the project. Project management would be carried out in concert with the steering group that, in fact, may call upon the management

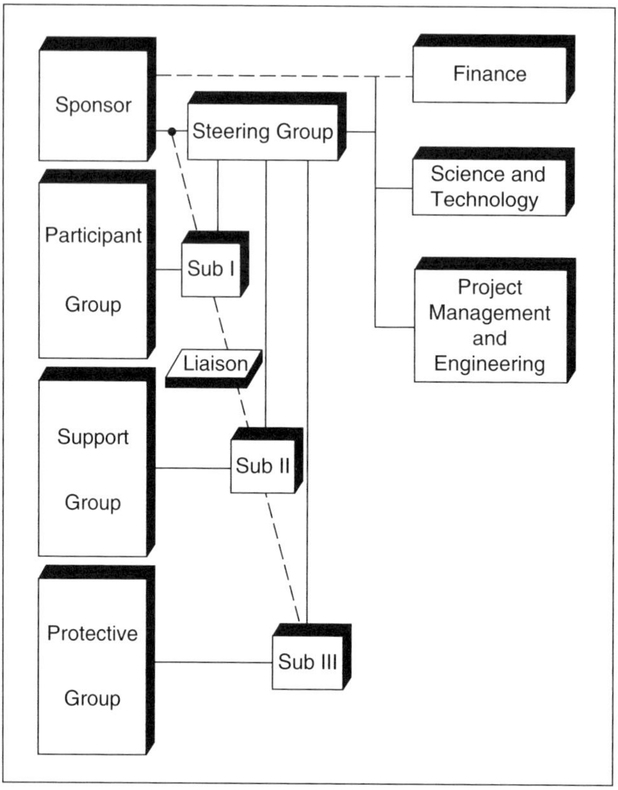

Figure 4.3 Links and liaison for group interaction. (From Warnock, 1978.)

team to perform special tasks or to change direction. The project management team, because of its close view of the undertaking, will be called on to assess risks and determine contingency plans, but the sponsor or the steerlng group retains the responsibility for accepting the risk.

4.2 ORGANIZATION

Tidal power developments are generally in the class of *giant* projects (Sykes, 1979), costing several billions of dollars and requiring substantial "proving-out" funds. Such funds may represent about 3 to 5 percent of the ultimate capital costs and range up to $100 million and more. In view of the magnitude of the costs required for the proving-out process, the investigation of an estuary for harnessing its tidal energy should be carried out in phases. This provides the opportunity, at the end of each phase, to decide to stop the investigation or to commit funds for further specific studies to a given schedule. As a result, the sponsor (or authority) maintains control of the study, and the required budgets and commitments can be limited to those steps necessary to provide sufficient information for the next decision point. Decision points should not only occur at the end of each phase but should also be built into a particular phase of the study should any reason or reasons emerge that would prejudice the true worth of the undertaking viewed in the proper, long-term context. Even after the expenditure of substantial sums, the investigation may have to be terminated.

4.3 FEASIBILITY STUDIES

An idea to exploit a resource becomes a project when one or more sponsors has spent sufficient money to demonstrate that a *prima facie* case exists for turning the idea into reality. This initial investigation should be the very first project overview in that it should consider all the main requirements of the project, which are apparent at that stage, to determine if, in principle, they can be met successfully. If an idea passes this initial screening, often referred to as the preliminary feasibility study, it should be considered to have a larger team assigned to it for further serious investigation, that is, the primary feasibility study. The latter study is a comprehensive review in appropriate detail to determine whether or not the proposed project is worth further, and major, consideration. It should examine, imaginatively, all key aspects requiring an experienced multidisciplinary team. This is not a study to be either rushed or skimped on since, on the strength of the results of such a review, increasingly large sums will be spent and risked. This first, thorough examination of a project should diminish the possibility of surprises by eliminating the avoidable ones at least and will have given full attention to optimizing the project. It may, perhaps, show that the project should be postponed until certain key events take place that the sponsors may or may not be able to influence. Assuming that a project survives the first intensive comprehensive study, the resulting document provides a blueprint on how to proceed and is continually expanded and updated until the project is accomplished or abandoned. This process will justify the necessary enthusiastic confidence that all matters critical to the project have been identified and the best solutions developed.

The preliminary feasibility phase may be preceded by initial phases dealing with activities to assess potentialities and general characteristics of possible developments and the physical data available and required. A good database is obviously essential for any subsequent studies such as for mathematical modeling of the tidal regime. The extent of effects of a tidal power project, and the difficulties posed by the environmental conditions under which some of the data must be obtained, makes data acquisition a relatively high-cost activity. The major categories of data collection are geophysical and geological surveys; wave measurements by buoys; tide gauging, including offshore gauges; current velocities; suspended and mobile bottom sediment concentrations and salinity; and surveys of waterfowl and shore birds (both migratory and indigenous), intertidal, sublittoral, and benthic fauna, and sites of special scientific or other interest. However, in setting up a data collection program, the first step is to collect and collate the available existing information that will provide guidance on the intensity of a data collection program for particular categories. Again, because of the relatively high costs of data collection in an ocean environment, the program should be phased to provide data adequate to the formulation of the project for the next decision point.

The preliminary feasibility or prefeasibility study of the energy potential of an estuary requires the determination of the likely cost of a project, the probable benefits of incorporating a varying energy source in the electrical utility system, and the possible impacts on humans and the environment. The main objective is to establish all of the basic factors that will enable a sound decision to be made whether or not a primary feasibility study, costing tens of millions of dollars, should proceed. This obviously requires a coherent and coordinated program of field investigations and studies, which would include sensitivity analyses of the costs of key factors. The specific objectives of the prefeasibility study would, therefore, include:

1. Preliminary definition of all schemes with performance and costs on a comparative basis and identification of environmental concerns applicable to each, resulting in the short-listing of the most promising schemes and sites (possibly up to three in number) for further examination
2. Selection of the best constructional approach for each short-listed option
3. Identification of the best way to integrate tidal-electric energy into the electrical utility system
4. Determination of any environmental concerns that could prohibit development

The organization for feasibility studies will depend not only on the infrastructure and facilities available in the country but also upon policy and political expediency. The organization of the Severn Estuary, United Kingdom, and Bay of Fundy, Canada, prefeasibility studies provide two examples.

Severn Estuary Studies The aims of this prefeasibility study were: (1) to narrow the options to one or two schemes or sites; (2) to select the best constructional approach for each option; (3) generally, to clarify the status of a tidal barrage among the renewable sources of energy; (4) to clarify the status of tidal power against conventional sources, such as coal and nuclear power, and to consider how it would fit into the electricity supply system; and (5) to determine whether there were any insurmountable environmental objections.

The overall program was carried out for the UK Department of Energy by a variety of organizations, including civil, mechanical, and electrical engineering contractors and consultants in industry, governmental laboratories, institutes, university departments, and the like. The extent of the environmental studies made full use of knowledge and expertise available in public and private institutions. The program was subdivided into a number of subject areas: data collection, mathematical modeling, engineering, economic analysis, evaluation of social and industrial impacts, and environmental impact assessment. Obviously, good coordination had to be a vital feature of such a program because of the highly interactive nature of the work, namely the results of some studies were required to provide refined input data for other study areas.

The investigation was carried out under the overall guidance of the Severn Barrage Committee, composed of 23 members drawn from the political, governmental, and private sectors. A Working Party on Tidal Power acted as the technical secretariat and the day-to-day management of the program was carried out by the Energy Technology Support Unit for the Department of Energy assisted by a firm of consulting engineers experienced in coastal and estuarial projects.

The Severn prefeasibility study was completed over a period of 2.5 years at a cost of about £2.3 million.

Bay of Fundy Studies The organization for the prefeasibility investigation of tidal power sites in the upper reaches of the Bay of Fundy took into account, of necessity, the jurisdictional aspects of natural resources in Canada. The federal/provincial agreement for the study set out the objective as follows:

> To provide a firm estimate of the cost of tidal energy in relation to its alternatives on which to base a decision to proceed further with detailed investigations and engineering design. (Terms of References, 1975)

In essence, this objective required the determination of three specific factors: a firm estimate of the cost of tidal energy, an estimate of the cost of energy from alternative sources, and a comparison between the two values.

To meet the objective, the program of investigations was divided into two stages. The first stage provided, through parametric comparison of prospective tidal power sites and schemes supported by preliminary systems analyses and market definition studies, an initial basis for comparing the *at-site* and *at-market* costs of tidal energy and the system costs of displaceable energy from alternative sources. These studies enabled the progressive screening of tidal sites to facilitate the selection of a manageable number of representative projects for further analysis. The second stage focused on the optimization of the most promising sites and included a more in-depth technical, economic, and financial assessment for comparison of system costs of energy *with* and *without* tidal power.

To facilitate the study, the program was separated into five main, but interdependent, task areas that required parallel consideration. These task areas were: (1) tidal power plant design, (2) tidal power generation (including tidal regime modeling), (3) market and systems analyses, supplies and transmission, (4) socioeconomic aspects, and (5) environmental aspects.

The federal/provincial agreement established a seven-member management committee, composed of senior engineers drawn from the pertinent agencies of the federal and provincial governments involved, to direct the study under the general policy guidance of a six-member review board of senior government officials. The committee contracted the work with a number of consulting firms. Its chairman was assisted in coordination and liaison activities by a study coordinator. The organizational framework, shown in Fig. 4.4, was tailored to the management and direction of the adopted program in order to meet the tight schedule specified in the agreement and provide the results under rapidly changing conditions.

To ensure the coordination of activities and full use of available expertise within governnent agencies, the management committee appointed subcommittees, one for each of the designated task areas. The general terms of reference were as follows: (1) maintain an overview of its designated task area with respect to the objectives of the review program, methodologies, execution, and interpretation of results; (2) advise the managenent committee upon questions

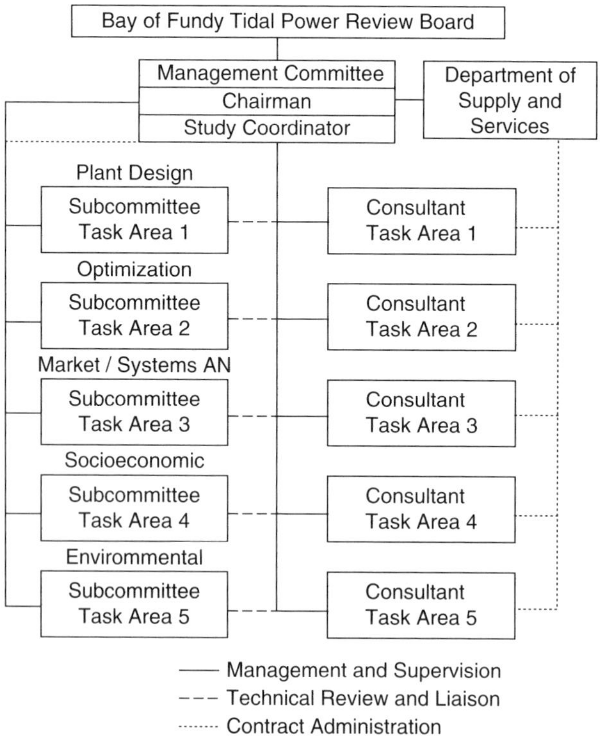

Figure 4.4 Organizational framework for Bay of Fundy prefeasibility studies (1975–1977).

specifically referred to it or others that may arise within its area of responsibility; (3) advise the study coordinator upon matters referred by him or her or upon other relevant matters; and (4) assist the study coordinator by providing him or her with copies of all relevant documents, findings, records of proceedings, or other relevant information generated by subcommittee activity.

The Fundy prefeasibility study was completed in 2 years at a cost of about $3.5 million.

CHAPTER 5

TIDAL POWER SCHEMES AND MODES OF OPERATION

How is it possible to harness this natural utterly renewable source of energy on a significant scale? Developing technology, given impetus by the Industrial Revolution, has been responsible for the emergence, in the last century, of large electric power and energy supply systems. Long-distance electrical transmission has freed tidal energy from the limitation of at-site consumption. Moreover, the extremely low and variable heads of a tidal cycle can now be efficiently exploited by low-head hydraulic turbines capable of generating and of pumping in both directions, making available four modes of operation in a development. Marked advances have also been made in the technology of construction in a marine environment and in the mathematical understanding of tidal cycle variations. The complex nature of the tidal energy problem has required the development of mathematical modeling techniques that, with the aid of high-speed computers, enable each area to be investigated at moderate cost and the energy yields and tidal effects of alternative development alignments and modes of operation predicted.

5.1 SINGLE-BASIN DEVELOPMENT, SINGLE-EFFECT MODE OF OPERATION

The simplest engineering solution to the recovery of tidal energy and its conversion to electricity would be to build a barrier across an estuary and equip

Elements of Tidal-Electric Engineering. By Robert H. Clark

it with turbines and with sluices to allow the basin thus formed to fill on the rising tide. At high tide, the sluice gates would be closed and the ebbing water would give rise to a head—a relatively small one—between its own level and that of the water of the basin. If the basin were emptied through the hydraulic turbines, energy of the tidal water would be retrieved, that is, single-effect operation on emptying the basin or *ebb generation*. This is the simple, tidal mill concept of Fig. 1.1.

Such a scheme, illustrated diagrammatically in Fig. 5.1*a*, is known as a *single-basin* development using *single-effect* turbines, or turbogenerators. As their name implies, single-effect generating units are designed to operate or generate power in one direction only. A sluiceway section, normally comprising a fairly large number of sluices, and a dyke section complete the dam or barrage and achieve the closure of the estuary.

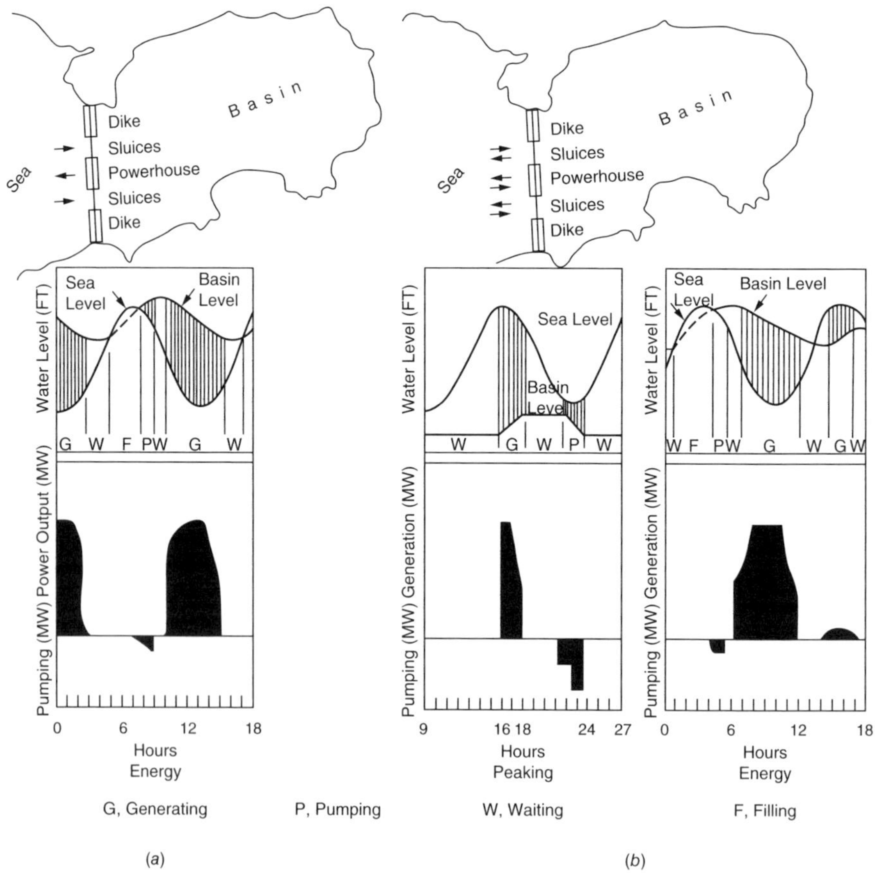

Figure 5.1 Characteristics of single-basin developments. (*a*) Single-basin, single-effect and (*b*) single-basin, double-effect examples.

5.1.1 Ebb Generation

Figure 5.2*a* illustrates in more detail the operation of an ebb generation scheme. At or shortly after the point of high tide when the basin is filled, the filling sluices are closed. After a waiting period, during which the water level in the ocean begins to fall, generation is started at a point in time where the head on the turbines has reached approximately half the value of the tidal range, that is, with the water level on the ocean side at approximately mean sea level. Operation proceeds with the flow from the basin to the ocean, past the point of low tide, or until the head on the turbines has reached the minimum value under which they are capable of developing power. At this point, the distributor gates to the turbines are closed and, after a short waiting period, the ocean level rises to what is now the basin level. When the two levels become equal, the sluice gates are opened and the basin is permitted to fill with the rising tide.

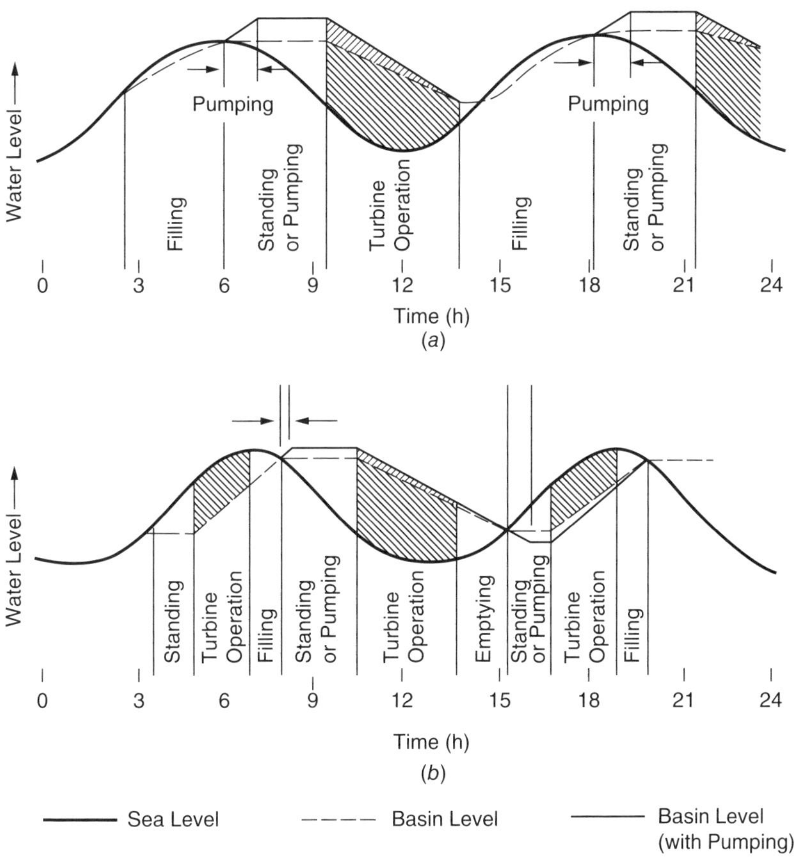

Figure 5.2 Water level versus time for ebb and two-way operation. (*a*) Single-effect operation on emptying and (*b*) double-effect operation.

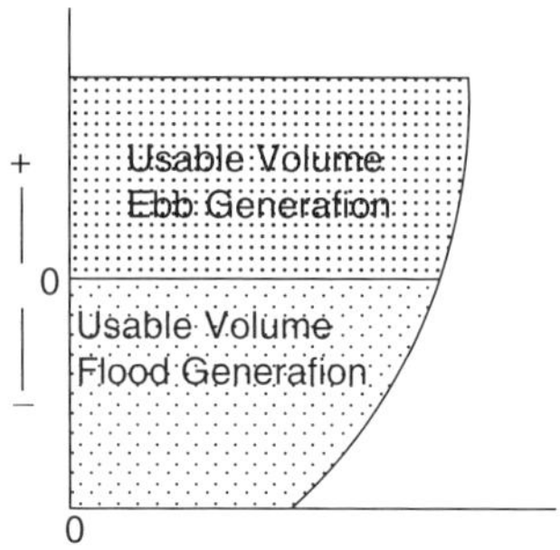

Figure 5.3 Basin area–elevation relation.

5.1.2 Flood Generation

By turning the powerhouse around so that its water inlet is on the sea side and its outlet on the basin side, energy can also be developed as the tide is rising and the basin is being filled—also a *single-effect* operation, but termed *flood generation*. However, a flood generation scheme would provide less energy than the equivalent *ebb generation* scheme. The primary reason for this is that the area of the basin or estuary decreases with depth giving rise to the following results: (1) the water levels in the basin would rise at a faster rate in the early stages of flood generation than they would fall during the same period under ebb generation; and (2) the volume of water that could pass through the turbines would be less for a flood generation scheme than an equivalent-sized, ebb generation scheme. These results are illustrated by the schematic basin area versus elevation curve of Fig. 5.3. Thus, generation on the ebb tide takes full advantage of the greater volume (and higher head) in the upper levels of the basin.

The civil works for the single-basin scheme are the least complex and hence tend to be the least costly. However, the single-basin concept of development with single-effect mode of operation produces energy only. It cannot produce continuous power and, therefore, cannot be considered a source of dependable peak power. The period of generation would be dictated by the times of the tides, which vary from day to day so that there would be many occurrences when maximum energy would be produced at times of low demand on the electric utility's system.

5.2 SINGLE-BASIN DEVELOPMENT, DOUBLE-EFFECT MODE OF OPERATION

With turbine technology now available, it is possible to use the same set of turbogenerators to produce electrical energy over both the rising and falling tides, that is, *double-effect* turbines could be installed. Such units are capable of generating, as well as pumping and acting as orifices, in both directions, that is, six modes of operation are possible. The attraction of such two-way generation is that it provides not only for the production of electricity during a greater portion of the solar day than single effect, but it also permits greater operational flexibility.

Referring to the point in the operating cycle of a single-effect scheme where the turbines have just reached their lowest head under which they can generate (Fig. 5.2*b*), the adjustable runner blades would be "feathered" so that the units would operate as orifices to assist the sluices in draining the basin to the level of the sea then prevailing. At this point the barrage is closed and, after a subsequent waiting period during which the ocean water level rises above that in the basin, the turbines are started in the generating mode and proceed to generate with the flow from the ocean into the basin. At the end of this generating cycle, the basin is filled and an ebb tide generating mode follows.

However, toward the end of both the ebb and flood generation periods during double-effect operation, the sluices must be opened sooner in order to reduce the basin level quickly to provide for the period of flood generation or to fill the basin as quickly as possible prior to the start of ebb operation. Thus, for any given numbers of turbines and sluices, slightly less energy would normally be produced by two-way or double-effect generation than by single-effect or ebb generation only. Under the former mode of operation, the operating heads are generally lower than for single-effect operation. On the other hand, the increased heads developed under higher tidal ranges could be more fully utilized with double-effect operation so that, as the tidal range increases, the latter operation could generate more energy per tide with the same number and size of sluices and turbines. Also, with two-way generation, the turbines would operate somewhat less efficiently at any given head than for single-effect or single-directional flow, since optimization for the latter condition must be compromised to obtain an optimum design for double-effect or two-way flow.

Although, as discussed in Chapter 11, there could be some capacity credit to an electric utility system from single-effect operation of a tidal-electric plant, double-effect operation could offer a considerably greater (two to three times) capacity support to a system because of its greater flexibility and could do so with only a nominal loss in energy production assuming that the plant would only be called upon to operate for capacity infrequently—on the order of 4 to 6 days per year—under a maximum capacity operating mode. It should be noted, however, that the principal limitation on the value of a tidal scheme as a capacity source is its inability to sustain capacity for more than a short time. However, under real system operation, the *full capacity* potential available from the tidal-electric plant might not be required at all times of emergency and a very wide range of operating strategies could be possible. Considering energy generation only, ebb generation would undoubtedly yield the minimum unit cost of energy.

It has been suggested (Grant, 1997) that for small-scale applications, a tidal barrage without gates allowing flow through the turbines from and into the basin might be attractive for unmanned operations in remote areas. However, without sluice gates, the differential head across the turbines would be less than that for a gated barrage. Such a simplified scheme could merit further investigation.

5.3 PUMPING TO AUGMENT TIDAL EFFECT

As already noted, the turbines can be designed to operate in the pumping mode under both single-effect and double-effect operation. The advantage of this capability would theoretically be an energy gain, if the pumping were to take place when the basin and sea levels were nearly the same and the water were used later in the tidal cycle when the head would be greater. The possibility that the energy output could thus be retimed by an hour or two might also be advantageous in the case of single-effect operation, the turbines could be started in the pumping mode at the point in the operation where the basin has just been filled by means of the sluiceways. Thus, under very low heads, the turbines operating in the pumping mode could fill the basin beyond the level of high tide, or *overfill* it (Fig. 5.2*a*). In the subsequent operation, the water that has been pumped into the basin at very low heads could be used for generation under much higher heads, a case where pumped storage would have an efficiency greater than unity.

Similarly, with double-effect operation, it would also be possible to *overempty* the basin before the reverse, flood-tide generating mode commenced. At the point where the rising ocean water level equals the declining basin water level, the turbines would be started in a pumping mode with flow from basin into the sea. Pumping would again proceed against fairly low heads and, during the subsequent flood generation, the additional space made available in the basin would be used under a considerably higher head (refer to Fig. 5.2).

A plant equipped with double-effect turbines becomes much more flexible with regard to the timing of its power production, so that it can be more closely adapted to the solar rhythm governing human activity. It would, in principle, be quite capable of producing dependable power at any desired time during the 24-h period. This is illustrated diagrammatically by Fig. 5.1*b*. However, although the plant can be operated to produce a dependable peak output, a substantial amount of its energy potential would have to be sacrificed in order to be ready for peaking production. In addition, energy would be required from the system to achieve the necessary pumping.

With its flexibility, the single-basin, double-effect development with pumping capability in both directions could, alternatively, be operated to provide maximum revenue. Generally, the value of energy in a power system will vary throughout the 24-h period. It would be possible to concentrate the energy production into the period of the day when high energy values prevail. An operating period of 14 days, for example, could be optimized so that the returns during this period, including the allowance for the cost of pumping energy, would be a maximum. However, such an operation could not provide for dependable peak production since peak production would have to be sacrificed, at times, for a greater energy production during another period of the day.

Although the addition of a pumping capability to turbogenerators would appear to be an attractive mode of operation, there are critical factors that must be examined before a decision can be made to include this capability in the design of the turbines. It has been shown (Bay of Fundy Tidal Power Review Board and Management Committee, 1977) that the beneficial effect of pumping is derived primarily under lower tidal ranges since the limited turbogenerator output for heads over the rated head deprives pumping of any significant attraction beyond a given tidal range (Cotillon, 1979). Moreover, turbines achieve their greatest efficiency when designed for a single mode of operation at a particular head of water. The wide range of operating heads for a tidal scheme would require some sacrifice of efficiency at the design point so as to achieve reasonably high efficiencies over the operating head ranges. If the turbine is designed to work in two directions, or to be able to pump, further concessions are necessary to achieve the best overall compromise in machine performance, and the cost of turbine control gear would be increased.

To summarize, although pumping at high-tide levels into the basin would appear to be an attractive mode of operation, it would have the following drawbacks (Severn Barrage Committee, 1981a): (1) there would, at best, be only a small energy gain compared with the output of machines designed for one-way generation only; (2) the timing of tides would rarely make this mode of operation convenient for the rest of the system; and, (3) the cost of the barrage would be increased and hence the benefit/cost ratio reduced.

In addition, although pumping at the La Rance (France) tidal power plant has been found to be attractive overall, since the mean tide there is approximately 8.5 m and the ratio of the neap-to-spring tidal ranges approaches 1:6, the Bay of Fundy Tidal Power Review Board and Management Committee (1977) pointed out that, in Minas Basin (Canada), the mean tide is 12 m and the ratio of neap-to-spring is only about 1:2. The latter's studies confirmed that pumping is most beneficial at low tidal ranges. Moreover, at the La Rance tidal-electric plant the rated head of the turbines is 5.52 m, and the lowest tides are of the order of 3 to 4 m; accordingly, the use of turbines as pumps at neap tides to increase operating heads during the succeeding generating period increases production not only as a function of the increased head but also by moving the operation of the turbines into the region of higher efficiency on the hill chart (refer to Fig. 9.8). The turbine rated heads in the Bay of Fundy (Canada), on the other hand, would be somewhat less than the ranges of the lowest tides that would allow the turbines to operate nearer to peak efficiency. Although pumping at the Fundy sites would increase the subsequent energy production as a function of head, there would be little effect from changes in turbine operating efficiency.

5.4 LINKED-BASIN DEVELOPMENTS

The basic characteristic of the single-basin scheme is that its output of electrical energy is discontinuous and tied to lunar timing with its semidiurnal and lunar cycles, unless substantial energy is sacrificed to adjust the timing of the output. As emphasized in Chapter 2, the lunar cycle is constantly drifting into and out of phase with the solar cycle, about every 2 weeks, and hence with industrial demands that are geared to the latter. This discontinuity of output posed a major problem early in the twentieth century when each local market was served by, and dependent on, a single source of supply. Obviously, continuity of output from that single source was of prime importance.

With the extremely large power systems that now exist, continuity of supply can be assured by the diversity of power sources feeding any given area. A new source is now generally assessed for: (1) the amount of power in kilowatts (kW) it can produce at the time of peak demand; (2) the amount of energy in kilowatt-hours (kWh) it can produce during a year; (3) its availability and reliability; and (4) environmental effects. However, the characteristics of a large power system tend to change with growth, so that a source with a continuous output may be needed at times to maintain a balance of generating facility characteristics within the system.

The simplest and least expensive, hydraulically linked, double-basin scheme is that attributed to Decoeur (1890) who suggested it for several locations in Europe in the 1890s. It was subsequently proposed for the International (Canada–United States) Passamaquoddy tidal power project (International Joint Commission, 1961), the development of several estuaries in the upper reaches of the Bay of Fundy (Department of Mines and Resources, Canada, 1945), and at other locations. The popularity of the Decoeur scheme probably lies in the simplicity and inexpensive installation relative to other double-basin arrangements as well as to the fact that it ensures continuous power generation.

The two basins are separated by barrages from the ocean and from each other. This development concept would benefit from a coastal configuration favorable to the creation of the two basins, such as in the case of Shepody Bay and Cumberland Basin at the head of the Bay of Fundy. As illustrated by Fig. 5.4, each basin would be connected by a set of sluices to the ocean. One set would serve to fill the *high* basin from the ocean and the other to drain the *low* basin to the ocean. The powerhouse would be situated in the cutoff dike between the basins or in the isthmus separating them, depending upon the topography. Generation would take place continuously and always in the direction from the *high* to the *low* basin, so that relatively simple generating equipment, viz. one-way hydraulic turbines, would be required.

From the point of view of power production, the smaller of the two basins would be chosen as the high basin. Figure 5.4 illustrates the operation of a linked-basin scheme. When the upper basin and sea level are at the same elevation, shortly after high tide, the sluices connnecting the upper basin to

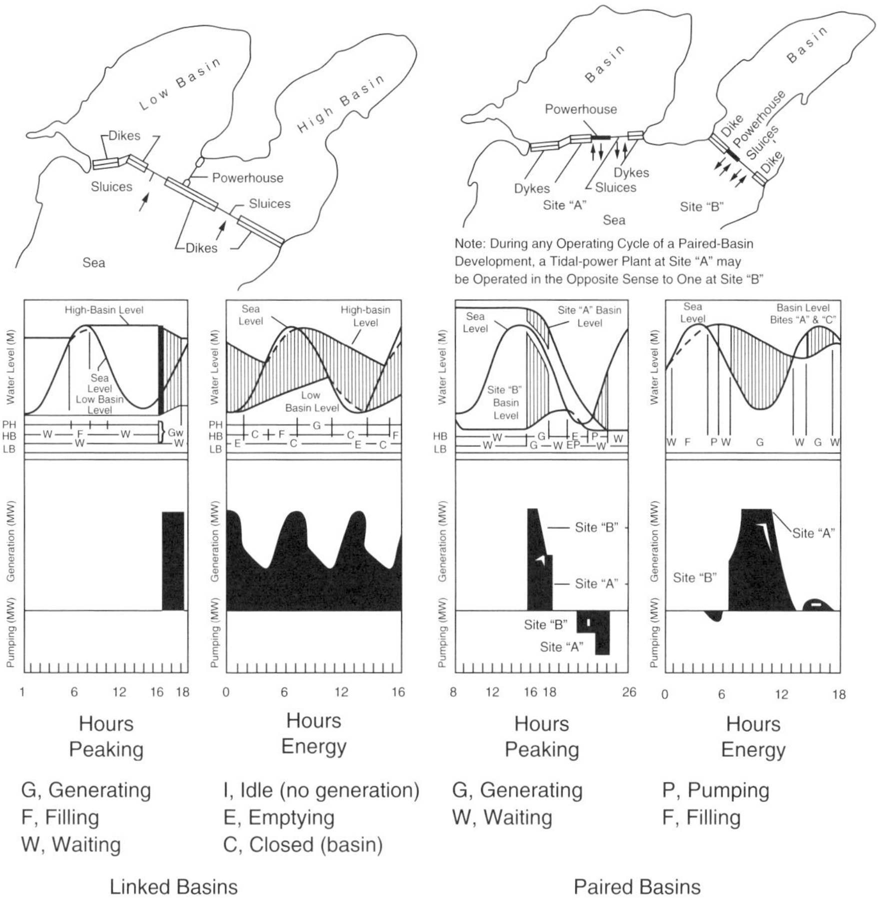

Figure 5.4 Characteristics of double-basin developments.

the ocean are closed. This basin is then emptied through the turbines into the lower basin, which is also closed to the ocean. When the rising level of the lower basin equals that of the falling level of the ocean the sluices connecting the lower basin to the ocean are opened so that the level in that basin falls with the ebb tide. Thus, the turbines now generate under the head between the upper basin and the ocean. At low water or ebb, the lower basin would again be cut off from the ocean and filled from the upper basin. When the falling level in the upper basin coincides with the ocean level rising on the flood tide, the sluices connecting the upper basin to the ocean are then opened and generation proceeds using the flow from the sea into the lower basin via the upper basin. When inflow ceases at the turn of the flood or high water, these gates are then closed and the cycle is repeated.

A double-basin scheme tends to be relatively higher in capital cost than a single-basin development since the former requires an interconnecting waterway in which the power plant is located, as well as dikes equipped with sluiceways for each basin to control their levels. Moreover, its total energy production would be lower than that from a single-basin scheme using either of the basins. It is obvious, therefore, that its energy output would have a higher unit cost than that of a single-basin scheme using either of the basins. On the other hand, there may be particular conditions related to geography or, perhaps, to power system demand to favor clearly this type of development.

This hydraulically linked, double-basin scheme can also be operated to produce dependable peak at a predetermined time of day as well as continuous, but variable, amounts of energy, even with one-way hydraulic turbines. The output of such operation is shown diagrammatically in Fig. 5.4. The production of dependable peak again entails some sacrifice of energy since both basins have to be held at their starting positions (maximum level for the high and minimum level for the low basin) to wait for the time of the dependable peak demand.

Bernstein (1965) describes several modes of operation of double-basin installations, some requiring turbines with pumping capability, others with additional turbine installations in each of the barrages separating the basins from the sea. Even schemes with three basins have been proposed to obtain a constant head regardless of the tidal phase. Such schemes have the objective of continuous output with greater utilization of the tidal energy potential at a site than that available by the simple Decoeur arrangement and mode of operation. However, the more extensive civil works and more sophisticated generating equipment required by such complex schemes result in a higher unit cost of energy.

A recent proposal put forward to reduce the civil costs of a linked, double-basin scheme relies on a different gate arrangement concept, combined with particular structural components, to allow the generation of energy from the high basin to either the low basin or the sea. This concept has been styled the "delta-connected tidal powerhouse" (Baker and van Walsum, 1983). In common with other schemes of increasing complexity, the objective of this concept is to reduce the higher civil costs associated with a linked, double-basin development while still maintaining its relatively minor advantage of continuous power output.

The single-basin, with either single-effect or double-effect turbines, offers the least unit cost of energy and can extract close to the maximum developable potential of the site. This conclusion, as suggested by Bernstein (1965) and supported by studies at various potential sites (Bay of Fundy Tidal Power Review Board and Management Committee, 1977; Severn Barrage Committee, 1981a), would not exclude the possibility that the simple, double-basin scheme may be better suited because of some specific utility system requirements or other considerations such as, for example, navigation.

5.5 PAIRED-BASIN DEVELOPMENTS

If operations of two single-basin developments were integrated and made complementary, the results would be a *paired-basin* development. Since flexibility of operation is the key attraction of this arrangement, the participating single-basin developments should be equipped for double-effect operation.

Paired basins could allow a continuous production of energy during the day by a judicious choice of water levels in the basins. However, such an operation, achieved at a cost of a reduction of energy output, would likely be undesirable in a large network. If they were operated for maximum energy production, the operation of each would be the same as if operated separately, provided other conditions of the load demand remained the same. The output would, therefore, be the sum of their energy production.

Actually, the main advantage of a paired-basin arrangement over an equivalent single-basin development would be in dependable peak operation. Ebb operation of one plant would produce an energy output pattern (with respect to time) that would be roughly a mirror image of that produced by flood operation of the other plant. Thus, when two single-basin double-effect plants are paired, one can be operated as a high basin and the other as a low basin, or both in the same mode as required, to obtain a more uniform combined-peak production. Typical outputs for peaking and energy operations of a paired-basin development are shown schematically in Fig. 5.4.

5.6 RETIMING OF TIDAL ENERGY

It is evident from the foregoing review of basic tidal power development schemes and modes of operation that the single-basin, single-effect scheme offers the lowest unit cost of energy and can develop the greatest proportion of the potential energy at a particular location. However, its output is available for only two periods of about 5 h each per day. Internal retiming or regulation through double-effect operation with pumping would not seem to be economically viable, and the use of multiple basins introduces further complexities and would likely add materially to the cost. There are several practical approaches to marketing raw tidal energy. Large, modern electric power systems are capable of absorbing energy from intermittent generating systems, particularly when that energy output is entirely predictable in magnitude and time, such as that from a tidal power plant. Should the system include hydroelectric generation, the output from a hydroelectric plant could be synchronized with that from a tidal-electric plant, or gas-turbine capacity could easily backup the tidal plant output. Some of the output from a tidal plant could be retimed either by using a pumped-storage plant (Atlantic Tidal Power Programming Board, 1969) or by means of a compressed-air energy storage (CAES) facility (Canadian Atlantic Power Group, 1988). These possibilities are discussed in Chapter 11.

CHAPTER 6

BASIC DATA

6.1 INTRODUCTION

A good database is essential for a prefeasibilty study if there is to be minimum risk that the results of the engineering and environmental studies of the final, short-listed sites will not be overturned except, perhaps, by some external factor, such as the price of fuel. Admittedly, for a reconnaissance-type study or preliminary assessment, some fairly broad assumptions may have to be made with regard to foundations, availability of materials, effects of sedimentation, and the like. For such assessments, there is little alternative since their purpose is to provide a reasonably informed basis for a decision whether or not it is worthwile to fund field investigations. Of course, for such assessments, all available physical data and information relative to the estuary should be collected and collated. On a cautionary note, however, it appears to be an inherent trait of engineers to minimize costs, that is, an optimistic bias. This trait has been evident in the estimation of risks and costs, particularly for new energy process plants (Merrow et al., 1979).

The major types of data necessary for prefeasibility studies are:

1. Bathymetric
2. Seabed conditions and geology

Elements of Tidal-Electric Engineering. By Robert H. Clark

3. Tidal levels and currents
4. Wave heights and periods
5. Suspended and mobile bottom sediment concentrations
6. Water quality
7. Ecosystem characteristics related to the natural regime of the estuary

6.2 GENERAL PHYSIOGRAPHY OF THE ESTUARY

The three-dimensional shape of the estuary, together with its general dimensions, provides information relative to its tidal resonance characteristics. Adequate bathymetric data in the vicinity of possible project sites and in those portions of the estuaries that would form tidal basins are essential, not only for project and construction cost estimates but also for estimates of the volume of water that can be stored and hence the potential energy available (Chapter 3). The volume of fresh water contributed to the estuary by its tributary streams should be determined since this may constitute a significant factor of the energy output and the operation of a tidal electric plant (Garrett, 1977; Douma, 1981).

The water depths can be obtained using an echo sounder that, when combined with tide curves, interpolated from recordings at gauges or nearby ports, produces a chart of bathymetry. For preliminary assessment studies, values of areas and volumes may be estimated from available charts and maps that could be supplemented using echo sounders or by a combination of aerial photography and ground control. Thus, at predetermined intervals of time during the flood and ebb of the tide, successive aerial photographs of the intertidal zone could be obtained over one-half of one tidal cycle. The line of contact between the foreshore and the water surface as indicated on the aerial photographs, with appropriate adjustments for variations in tidal level, would enable the establishment of precise contours in the intertidal zones. Correlation of these contours with the hydrographic and topographic contours would permit the measurement of basin areas at any desired elevation.

Because of the potential impact of changes in water levels on existing works around the periphery of the tidal basins under consideration, field data should be obtained with respect to the elevations of the crest of existing dikes, elevations at low points of adjacent highways, roads, and railroads, and with respect to the controlling elevations of all other man-made structures. On the basis of the volume–elevation and area–elevation relationships for each of the basins under consideration and an evaluation of the controlling elevations associated with existing works around the periphery of a basin, the practical upper and lower limiting levels, which would have little effect on existing structures or could be readily corrected at nominal cost, could be established.

6.3 GEOLOGY

The geologic formations in an estuary are of considerable importance since it is highly desirable that the site locations be free from active or, indeed, major but inactive faults because of the potential hazard to tidal power structures or possible increased costs for remedial treatment. Foundations are of primary importance because of the massive nature of the structures in a tidal power development. The foundations must be adequate in compressive strength, water tightness, and other properties. They should also be free, as far as possible, from unconsolidated materials that may undergo consolidation during the course of construction operations, or require removal with replacement by other materials. Also important to the economics of construction are the sources of fine and coarse aggregates, durable rockfill, and suitable armorstone that, hopefully, can be located close to the site.

A relatively straightforward method of gaining an adequately comprehensive knowledge of the unconsolidated sediments between seabed bedrock and of the underlying bedrock would be by the use of various types of drill rigs. Such a method would, however, be quite costly in deep tidal waters.

A geophysical survey can provide a picture of the regional geology of an estuary, information on the conditions of the seabed likely to affect the siting of a barrage, as well as the distribution of material suitable for fill. The estuary would be first marked off by a square grid of lines whose spacing would depend upon the accuracy of data required, for example, the lines might be chosen in the order of 2 km apart. Echo sounding (e.g., by a Dual Frequency Raytheon Sounder, DE-719, 200 kHz, and 3.5 kHz, respectively), sidescan sonar (e.g., 200 kHz Klein Side Scan Sonar Model 401), boomer and pinger continuous profiling systems would be used along these lines to build up a picture of the depth to seabed (bathymetry), an acoustic picture of the seabed over a fairly wide swath, and sub-bottom profiles with penetrations up to about 25 m. Under favorable weather conditions, a large area of the estuary bed can be covered in a relatively short time and at fairly low cost.

Seabed samples can be taken periodically to assist in specific problems of interpretation. The soft sediments at the top of the bed can be sampled by the "Shipek" grab, which is a spring-loaded scoop. Deeper soft sediments may be sampled by a vibracorer, which is a sample tube driven into the ground by means of a weight and motor driven vibrator. This method enables the recovery of relatively undisturbed samples from the full profile of unconsolidated sediments between seabed and underlying bedrock, and at a reasonable cost. Core recoveries depend to a large extent on the type of materials being sampled and may be as long as 10 m. Some compaction and remolding may result from the sampling procedure, but the sediment core should, in general, prove adequate for standard laboratory strength and consolidation tests. Depending on the strength of the tidal currents, the time "windows" for the operation may be severely limited—perhaps to one hour during each slack-tide period.

In order to provide necessary data on bedrock structure for correlation with geophysical profiles in the vicinity of the center lines of the prospective sites, diamond drilling can be carried out in the intertidal zone at substantially less cost than from floating equipment. Each drill hole in the intertidal zone should be located as far from the high water mark as possible in order to overlap with the ends of the geophysical profiles. Of course, an important consideration is the positioning of each hole so that there is a reasonable period of time during which the drilling can be carried out.

Accurate position fixing must be established. A navigation system should be established with shore stations precisely surveyed and spaced so that the boat can interrogate two stations at a time. The distances between the boat and the stations can be determined every few seconds (e.g., a range/range Mini Ranger Mark II system), and a minicomputer can automatically calculate and plot the grid coordinates of the vessel. The accuracy of positioning of the vessel should be quite high, although that of the geophysical survey line would be somewhat less accurate since the profiles would be obtained by an instrument towed behind the vessel and so could also be affected by cross-tidal flows affecting the distance between the instrument and the ship.

The results of the geophysical survey should be plotted on maps of about 1:50,000 scale showing the bathymetry at contour intervals of not more than about 5 m and the thickness of surficial sediments with contour intervals of about half the bathymetry contours. A map showing "depth to bedrock" may also be useful.

The *seismicity* of the estuary region should be assessed in order to provide estimates of ground accelerations and associated return periods.

6.4 TIDES

Since the energy varies with the square of the tidal range, the tides of an estuary are obviously its most significant feature, otherwise it would not be a subject of investigation. Knowledge of the tides necessary for safe navigation has usually been the prime consideration in the collection and prediction of tidal data and in the preparation of hydrographic charts. Nevertheless, continuous tidal level measurements in an estuary with sufficiently high ranges for energy development are generally not available.

In highly developed and settled regions in which an estuary may lie, at least one permanent tidal gauge has usually been established. Tidal levels may also have been measured for short periods of time at specific locations in the estuary for particular purposes. Where such water level data are available, the gauges have usually been operated by port, water, or local authorities for short periods at different times over the years and cannot be compared directly.

Correlated with the records at the permanent gauge, data from temporary gauges may provide an adequate basis for the evaluation of tidal power possibilities for the estuary. Analysis of the amplitude and phase of the tide from the records at a temporary gauge location will yield values of the harmonic constituents for a site that can be used to describe the tide. These values, known as the harmonic constants for a site, relate the tides to the motion of the sun and moon relative to Earth and are used for tidal prediction. The values of the harmonic constants at a site should remain the same from year to year provided the effects of known long-period astronomical variations are included in the analysis.

Although continuous data over a 12-month period provides the best basis for analysis, a minimum 4-week continuous period of tidal data can provide a reliable harmonic analysis. The vertical tide at any place may be expressed in terms of a simple harmonic series as

$$h = h_0 + \sum_{n=0}^{\infty} h_n \cos(\omega_n t - \alpha_n) \tag{6.1}$$

where h_0 = height of mean water level relative to datum
h_n = nth harmonic constituent
ω_n = angular velocity
α_n = phase displacement
t = time

Equation (6.1) indicates that at most places the tide may be represented by a sum of sinusoids of known frequencies but whose amplitudes and phases are characteristic of the site. A harmonic analysis determines the amplitude and the phase of each harmonic component or constituent from observations on the local variations of water level. The various constituents can be related to the tide-generating forces, which are known from the astronomical relationships among Earth, sun, and moon, as well as interactions among the constituents themselves. The number of constituents definable (i.e., their amplitudes and phases) is governed by the length of the data record. Many of the constituents are very nearly identical in frequency and, therefore, require an extremely long record for their resolution. The number of constituents, n, is very large—over 500—but, in practice, a few of the harmonics in the diurnal and semidiurnal bands aggregate much larger than all others combined and suffice to give a good approximation of the tide (refer to the Glossary for definitions of the principal tidal constituents). In Cobequid Bay, Canada, for example, the semidiurnal constituents have the following values: M_2 is 606 cm, L_2 and S_s approach 100 cm, and several others are in the 20- to 40-cm range. The only significant diurnal constituents are O_1 and K_1, which are about 13 and 17 cm, respectively. Since tidal energy is proportional to the square of the amplitude, the M_2 constituent represents about 95 precent of the total energy, and the use of six constituents accounts for over 99 percent of the energy spectrum.

The reconstitution of the tides at various prospective sites provides an insight to the distribution of the tidal ranges during a year as well as to their characteristics (Godin, 1980). In addition, simultaneous tidal observations over at least a month enables a comprehensive picture of tidal propagation throughout the estuary under consideration to be assembled.

In order to obtain a high-quality database of tidal observations in the offshore areas of an estuary for input and calibration for the numerical modeling of the area, a tidal measurement program is required during the initial phases of an investigation. Shore-based observations are not adequate to provide a comprehensive understanding of the physics of the estuary or to enable more accurate predictions of the impacts of changes to the system by the construction and operation of a tidal-electric plant. The reasons for the generally inadequate tidal data are obvious; large tides often imply foreshores that are dry for kilometers and such tidal ranges tax most instruments.

Measurements of water levels in offshore areas must, therefore, be undertaken but, as in any program of oceanographic observations, there are risks with respect to the recovery of moored instrumentation from various factors, such as the fishing industry, weather, corrosion, navigation, as well as the failure of key electronic components in the mooring hardware. The instrument package should, therefore, be simple, uncluttered, and as unobtrusive as possible, particularly if it is moored in a region with a large offshore fishing industry.

DeWolfe (1979) describes such an instrumentation package designed for the Bay of Fundy (Canada) two-part tidal observation program during 1976. The configuration of the mooring is depicted in Fig. 6.1. The package is uncluttered with ground tackle. The pressure gauge (in this case an Aanderaa TG-3A) is set in a hole in the concrete block anchor and is secured to the top of the release (AMF Model 325) with a slack lanyard. The release, complete with flotation collars of syntactic foam, which is noncompressible, is shackled to a stainless steel rod that protrudes from the anchor. An OAR submersible flasher and an OAR submersible radio beacon, complete with pressure switches, are secured to the flotation assembly, projecting downward. The pressure gauge is secured in the anchor by packing bread between it and the anchor. This provides rigidity during deployment of the package but ease of recovery since the bread dissolves during the duration of the mooring. The instrumentation is lowered to the bottom by a "drop-hook," which releases the package immediately on contact with the bottom; the resulting mooring is about $0.75\,m^2$ by 1.5 m high. To recover the package, the release is actuated by an acoustic command from a ship-borne AMF 200 deck unit. As the package starts to rise, the pressure gauge produces a turning movement that causes the release to rotate 180° and surface upside down. Once surfaced, both the radio beacon and light enable easy location; the release hydrophone remains submerged for acoustic ranging on the package. The anchor is, of course, expendable.

The first part of the tidal observation program was designed to provide minimum coverage for calibration of the model and the second part, to augment the database where analysis indicated a need for more information and to verify the model. The inshore gauges were left in place for about 3 months and the offshore packages for not less than 2 months. The deployment

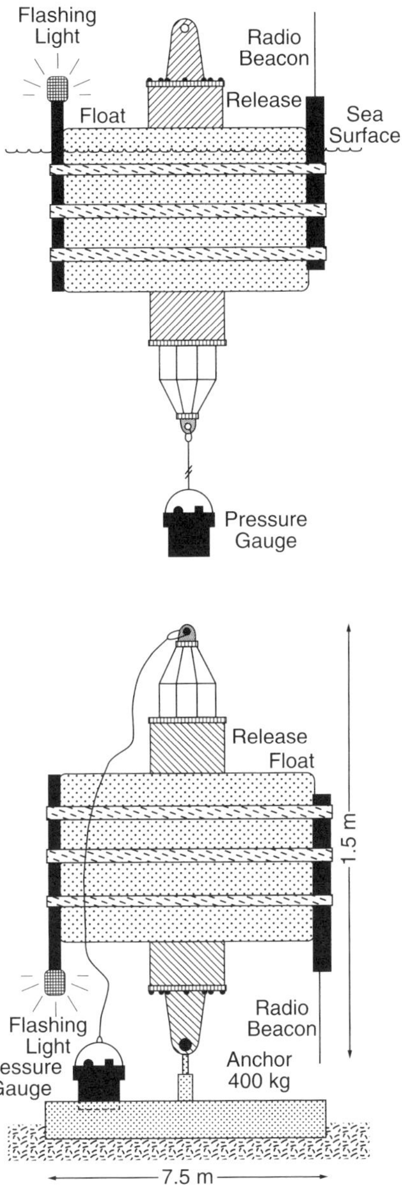

Figure 6.1 Configuration of instrument mooring for offshore tidal observations. (From Amos, 1977.)

of the gauges is indicated in Fig. 7.3. Overall, the percentage of good data for the project was about 71 percent; the 29 percent loss was caused by loss of mooring and instrument failure.

The Aanderaa tide gauge records, on magnetic tape, the hydrostatic pressure above the instrument at half-hourly intervals. The pressure resolution is 0.5 mg (equivalent to ±5 mm water) and a time-keeping accuracy of ±3 s/month, permitting a high degree of confidence in the quality of data. Analysis of the time series of pressure data provides the amplitudes and phases of the various tidal constituents.

6.5 WAVES

Several aspects of any tidal power scheme are sensitive to wave climate, such as embankment protection, forces on gates and caissons, barrage elevation, caisson handling, turbine performance, delivery, and placing of materials by sea. Continuous wave data can be compiled by the deployment and maintenance of wave-rider buoys that may prove a frustrating experience, particularly during storms, which, of course, are of particular interest.

Wind data are required to establish the design criteria for waves and the amount of wind setup for the design of structures. It is necessary to have the data to forecast the height and period of extreme waves, that is, waves likely to occur in storms of such severity as might occur at average intervals of 50 years or more; some features of barrage design may require lower risks and therefore longer return periods.

Sites selected for the investigation of tidal power may be surrounded by a variety of topographic and hydrographic features giving rise to complex wind and wave characteristics. Winds are substantially influenced by the proximity of high topographical features, and wind-generated waves are influenced to a high degree by irregular shorelines, relatively long and narrow estuaries, extensive drying foreshores, islands, and shallows. Practically all wind-recording stations are land based and may be at considerable distance from the shoreline. In an investigation of the effects of waves on the design and operation of a tidal-electric plant overwater wind velocities are of primary interest. To obtain such velocities, winds recorded at land-based stations should generally be increased to simulate overwater velocities.

If no permanent wind-recording installation is available in the region of the estuary that can provide useful information for the preparation of wind records and analyses of velocities and return periods, then anemometers must be established to commence the collection of such data, particularly to supplement the wave data being collected. However, short periods of wind records can be extremely misleading with regard to return periods unless the limited data can be supplemented by special analysis by a meteorologist on the wind climate of the region.

6.6 TIDAL CURRENTS

The process of proving numerical models requires friction factors to be adjusted so that tidal propagation in the model matches that in the estuary. Moreover, a thorough knowledge of tidal currents in the vicinity of the prospective site and proposed construction marshaling areas is required for design purposes, as well as full information on tidal currents in the estuary for the construction and verification of a hydraulic model.

The monitoring of tidal currents should be carried out over periods of about 2 to 8 weeks to measure lunar variations in current speed and direction, temperature, and salinity. The site may also be equipped with a bank of such meters spaced at depth intervals to determine the vertical current velocity distribution over a lunar cycle and to assess the effects of bottom friction on tidal flow.

High-altitude aerial photography applied to the technique of water current measurement can provide reasonably good surface-current data.

6.7 SUSPENDED AND MOBILE SEDIMENTS

The understanding of existing sediment behavior in an estuary is a key issue in the study of a tidal power scheme in order to be able to answer the question: Will sedimentation significantly affect the operation of a proposed tidal power scheme in the estuary? (Amos et al., 1994). Large tidal estuaries revealing complex hydrodynamics and sediments require the collection of a great deal of field data as well as an understanding of basic sediment transport mechanisms relating to the intertidal zone and the formation of turbidity.

Realistic predictions on the rate and pattern of siltation resulting from the construction and operation of a tidal-electric plant require a knowledge of the existing regime. Therefore, it is necessary to amass the relevant data to answer the questions: What are the existing hydrodynamic conditions and their relation to the sedimentation character of the region? Can hydrodynamic predictions of changes (both spatial and temporal) resulting from barrage construction be developed? To answer the latter question, the rates of transport and instantaneous fluxes of material moving both in suspension and along the bed, and the measurements of the dominant physical and meteorological conditions considered to govern the character of the sediments and temporal variations (seasonal, lunar, and tidal) of these rates and measurements, must be available.

Generally speaking, the nature of siltation resulting from construction or placing obstructions in the tidal flow are a function of: (1) the degree of flow reduction caused by the construction—a tidal-electric barrage will be permeable, removing tidal energy during its operation with the result that the tidal flow will be much reduced over large portions of the tidal cycle so as to induce siltation; (2) the availability of material—siltation can occur only if sediment of the appropriate size is available; and (3) the continued supply of sediment material to the barrage site—the processes controlling the postbarrage origin and transport of sediment must be known to continue, or dwindle, or increase, in order to make the necessary projections based on sediment supply (Amos, 1979).

Samples to determine the distribution of bottom sediments should be collected on a predetermined grid of 1 to 3 km. A 0.2 m^3 Van Veen sampler would provide adequate subtidal samples; intertidal samples can be collected manually. The samples would be analyzed for grain-size distribution, carbon/nitrogen ratios, and macrofaunal assemblages.

Sediments in suspension can be considered to undergo three basic types of motion within the water column. First, there are the vertical motions induced by gravity (downward motion) and turbulence (upward and downward); effects of turbulence can keep very fine sand in suspension over the entire tidal cycle. Second, the sediment moves in a bipolar fashion within the tidal prism, traveling distances as great as 40 km/tidal cycle as, for example, in Minas Basin of the Bay of Fundy. The third type of motion of the sediment is the net residual movement due to the foregoing processes. It is this latter movement that is considered of primary importance to an understanding of the depositional propensities of the suspended load. Figure 6.2*a* shows the results of temporal fluctuations in suspended sediment fluxes with respect to time and depth at one location (east of Minas Channel) of an extensive water-monitoring program in Minas Basin (Amos, 1977). The results for each station monitored by the program were analyzed for the residual movements of suspended sediment, calculated by integrating the fluxes for each stage of the tide with respect to depth, and the results plotted (refer to Fig. 6.2*b*). These data indicate that the highest transport rates of sediment occur during the midflood and midebb stages of the tide. For the case shown, there is a residual movement of 2.7 metric tonnes/tide of sediment, landward across a water column 1 m wide.

The distribution of suspended particulate matter can be measured areally by use of the Landsat MSS (multispectral scanner) and vertically and temporally by direct sampling. A system for quantitative measurement of suspended sediment concentration from Landsat MSS is now operational. Based on chromaticity analysis, the system permits the surface calibration data of one satellite image to be applied to another by an adjustment for variations in atmosphere, water surface geometry, and solar angle (Munday, Afoldi, and Amos, 1980).

Data on bathymetry, sources of material and volumes of sediment added or removed from the system, suspended sediment concentration, critical erosion velocities, critical deposition velocities, rates of erosion, rates of deposition, flocculation rates, settling rates, and the appropriate sediment entrainment equations provide input and boundary conditions to the tidal regime model. The bottom distribution of sediments would be used to calibrate the model. Once calibrated, the model could then assess the postbarrage transport and deposition of material within the system.

One technique for the development of an answer to the basic question—Will sedimentation significantly affect the operation of a proposed tidal power scheme?—is to approach the problem in three steps. The first step would be to define the total volume of sediment potentially available for sedimentation

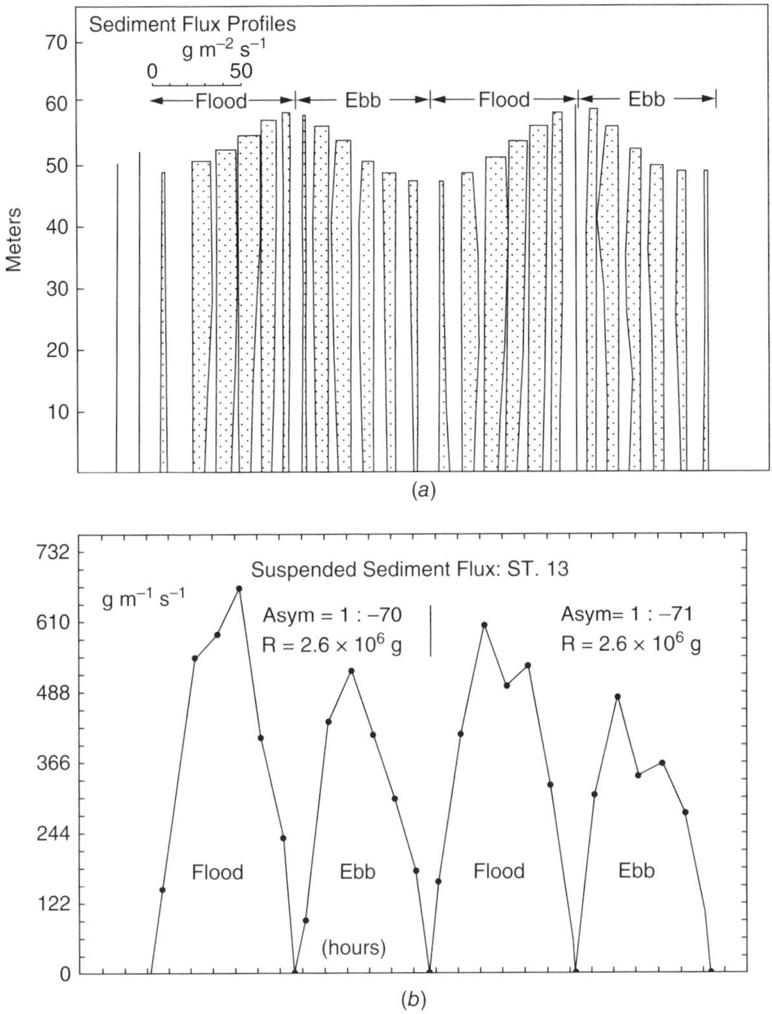

Figure 6.2 Suspended sediment transport at a station in Minas Basin. (From Amos, 1977.)

in the estuary over the useful life of the power development, such as riverine input, local bluff erosion, and the suspended sediment flux at the proposed tidal barrage site. Knowing the hypsometry of the tidal plant headpond, specifically the sublittoral and littoral volumes, it will be possible to determine the percentage reduction in the volume of the headpond assuming, as a worst-case scenario, that all available sediment would be deposited in the headpond. If the estimates show that postbarrage sedimentation could be significant, further, more complex, analyses would be necessary in order to define more precisely the volumes of sediment to pass into the estuary with the tidal plant in operation based on simulation of the system hydrodynamics and associated sedimentation responses. Therefore, the second step would be to examine the sediment fluxes into the headpond area based on numerical simulations of tidal and wave dynamics of the region, coupled with a three-dimensional sedimentation model and a detailed definition of the location and magnitude of the sediment influxes. Model output could be calibrated against satellite maps of suspended sediment in the estuary. The calibrated model could then be used to define long-term (e.g., 30-year) fluxes of sediment volumes to the estuary in the presence of a tidal barrage. The third step, if necessary, would involve the execution of the three-dimentional model over the life cycle of the tidal plant so as to account for the nonlinearities in the headward fluxes of sediment due to changes in bathymetry as a result of deposition. The result would be the best estimate of headpond sedimentation as well as providing indications of spatial variation in deposition in the headpond.

6.8 ECOSYSTEM CHARACTERISTICS

A discussion of the ecosystems that may be affected and the type of baseline data necessary to evaluate the impact of a tidal-electric plant is presented in Chapter 14.

CHAPTER 7

HYDRAULIC AND NUMERICAL MODELS IN FEASIBILITY INVESTIGATIONS

7.1 INTRODUCTION

Models or simulations are essential tools in tidal-electric engineering in order to predict or calculate the response of existing or hypothetical systems to proposed changes. As a scaled-down physical version of the *natural* system, it is known as an *hydraulic* model. Such a model is often seen as a miniature version of the real estuary with water running through channels and around islands, but it is still an analog simulation and only an engineering tool to solve engineering problems. As a numerical approximation through a system of mathematical equations to describe the prototype or natural system, it is known as a *numerical* or *mathematical* model (Ippen, 1966).

These two modeling techniques offer capabilities for specific purposes and complement each other in several ways. Their combined use as a *hybrid* model embraces the advantages of both, which is particularly important in the modeling of large estuaries, such as the Bay of Fundy (Canada). The Canadian Hydraulics Centre of the National Research Council of Canada conceived and developed the idea of the *hybrid* model, that is, a physical model of a certain region dynamically coupled to a numerical model of an adjacent region (Prandle et al., 1980).

It should be kept in mind that modeling of hydrodynamic phenomena requires that the many choices, which must inevitably be made prior to the design of a model, are based on reasonable compromises important to a

Elements of Tidal-Electric Engineering. By Robert H. Clark

specific investigation. All models must undergo a calibration phase during which they are adjusted to match a set of prototype measurements. Thus, a model can only be as good as its calibration and a calibration can only be as good as the prototype measurements. A useful discussion of modeling requirements is provided by Funke et al. (1979).

In general, hydraulic models are better able to model small, prototype areas where the flow is three dimensionally complex and nonlinear effects are significant. On the other hand, numerical models are better suited for large areas where good mathematical descriptions of the natural phenomenon under study are available (Crookshank, 1984).

7.2 HYDRAULIC MODELS

Hydraulic models are popular since they have an immediate appeal to an observer and can inspire considerable confidence. They provide an excellent visual representation of the fluid dynamic processes and are also capable of simulating nonlinear flow situations.

The validity of the hydraulic processes depends to a great extent on the scaling law chosen and on its relation to the physical variables under investigation. Keulegan (1966) provides a description of various methods by which model similarity laws can be obtained. In some cases, however, the validity of the model is so limited that quantitative results are unreliable, although qualitative observation may still be useful. For example, despite the wealth of experience accumulated, appropriate scaling laws remain uncertain for modeling the suspension, transportation, and sedimentation of materials. Generally, however, the smaller the model scale factors, namely the larger the model, the greater will be the model validity for nonlinear phenomena.

Thus, the hydraulic model is the better tool in those situations where there is much stratification, complicated boundaries and flow patterns, significant bottom effects, scouring or sedimentation problems, and random or short-crested or breaking waves. It is also possible to simulate Coriolis forces on physical models by means of a large-diameter, slowly rotating table (Gibrat, 1966). However, the hydraulic model of an estuary would have to be severely scaled down in size since there are major constraints on the diameter to which a rotating table could be fabricated. As a result, the benefits derived from rotating the model are offset by a general loss in simulation accuracy for tidal phenomena. On the other hand, numerical models can take into account, quite adequately, not only the Coriolis but also the shear and tidal forces, so that the turntable physical model is not a practical approach for large estuaries.

A number of hydraulic model studies of components of a tidal barrage will no doubt be required at scales varying from 1:20 to 1:60 to assist in the effective construction and operation of a tidal-electric plant (refer to Fig. 7.1). Boundary conditions for such modeling may be acquired from a

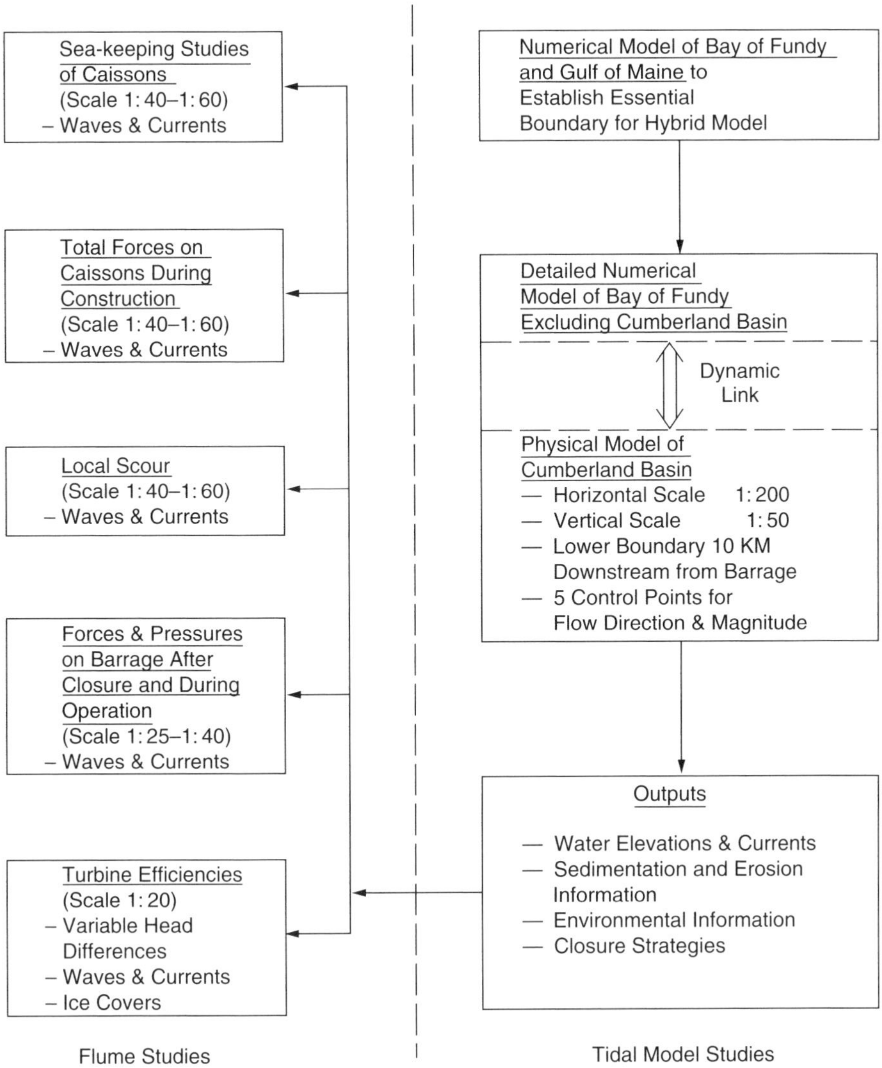

Figure 7.1 Typical problems requiring model studies. (From: Funke et al., 1979.)

low-distortion, hydraulic model of the estuary combined with a mathematical tidal model, a combination known as the *hybrid* model.

Some of the problems to be examined through hydraulic modeling would include:

1. Sea-keeping characteristics of turbines and sluiceway caissons during towing and sinking operations
2. Forces in tow lines during transportation of caissons

3. Total forces acting on the caissons after placement due to currents, waves, and tidal-head differences; special attention must be given to closure conditions, particularly to the end caisson when most of the barrage has been constructed and currents are severe through the remaining narrow opening to the basin
4. The local scouring in the closing gap due to intensified velocities
5. Investigations into caisson foundations
6. Oscillatory loading of the caissons under wave actions and its effect on the foundation
7. Strategies for end-dumping, especially during closure
8. Forces and pressure distributions acting on various structures during actual operation of the plant
9. Turbine efficiency as a function of normal and abnormal variables such as wave action or ice
10. Examination of sluice sizes and discharge coefficients

Many other problems to be examined by hydraulic models would likely emerge during the course of the overall feasibility investigation, but hydraulic model studies should not normally be undertaken until after the results of the prefeasibility investigation have shown the project to be distinctly viable. Undoubtedly, however, a wave flume would have to be available and equipped not only with a random wave-making facility to match the wave characteristics encountered in the estuary but also with a large-capacity, bidirectional variable-flow pump with water elevation control at both ends.

7.3 NUMERICAL MODELS FOR ESTUARIES

The numerical model is formed by a set of equations that describes the motion of a fluid in the presence of a multitude of forces. These continuous equations are then transformed into numerical approximations that result in a new set of equations of discrete time and space variables which may be numerically evaluated on computers. The outputs can be displayed in terms of tables of numbers or graphs, or even by three-dimensional visual representations of the output variables (Serrer and Crookshank, 1995).

Initially, numerical models were the simple one-dimensional (1D) type used to simulate the dynamics of a river system where variations across the river are small compared to variations along the river. They can be used for initial, broad scanning of different layouts and locations and operational conditions, tides, and the like. However, accompanying the advances in computer technology, two-dimensional (2D) modeling has reached a mature level. The current ability to model predominantly two-dimensional, mesoscale circulation is quite good, but three-dimensional (3D) flow modeling is still in the developmental stage since comprehensive and extensive field data sets are necessary for assessment of the simulation (Westerink and Gray, 1991). A new technique, the *smoothed particle hydrodynamic* (SPH) method has been extended to deal with free-surface incompressible flows. It is stated to be simple, robust, easy to program and to provide a reasonably accurate technique for complex problems, such as dam breaks, tidal bores, and the propagation of waves toward a beach (Monaghan, 1994). However, application of this technique to tidal-electric engineering problems will require further study and development and more powerful computers in view of the computational intensive solutions.

7.3.1 Numerical Methods

The validity of a numerical model depends upon three factors: (1) the adequacy of basic mathematical representation from which the model has been derived; (2) the proximity of the discrete time and space variables; and (3) the stability of the numerical scheme that is being employed. The numerical model is particularly suitable for simulating physical processes associated with time-varying free-surface flows along coasts and in estuaries generated by winds, tides, storm surges, floods, and dam and dike failures. Wind shear and Coriolis forces can easily be introduced into the numerical model. New generation software packages, such as Telemac-2D (Laboratoire d'Hydraulique de France, 1992), which is a two-dimensional model based on the finite-element method, are available to carry out these simulations.

The estuary dynamics of the mathematical model may be expressed by the following equations of continuity and motion, displayed in usual nonlinear form:

X motion:

$$\frac{\delta U}{\delta t}+g\frac{\delta Z}{\delta x}+f\frac{U(U^2+V^2)^{1/2}}{D+Z}-\Omega V=0 \tag{7.1}$$

Y motion:

$$\frac{\delta V}{\delta t}+g\frac{\delta Z}{\delta y}+f\frac{V(U^2+V^2)^{1/2}}{D+Z}-\Omega U=0 \tag{7.2}$$

Continuity:

$$\frac{\delta Z}{\delta t}+\frac{\delta}{\delta x}[U(D+Z)]+\frac{\delta}{\delta y}[V(D+Z)]=0 \tag{7.3}$$

where U and V are the velocities in the X and Y directions, respectively, D is the depth below mean water, Z is the tidal elevation about the mean, f is the friction term, and Ω is the Coriolis parameter.

The equations of motion include real depth, quadratic friction, and Coriolis acceleration but do not include advective terms that may be required to simulate strong eddy currents in specific areas of the model. The techniques used to solve the governing equations by numerical means fall into two broad categories: finite-difference method or finite-element method. The former method, at least until recently, has been the most popular.

Finite-Difference Method The basic equations are transformed directly into finite-difference equations by replacing the partial derivatives with finite-difference approximations utilizing a differencing technique such as forward or central differencing. For two-dimensional models, the spatial area under study is divided by a grid and the time domain is divided into field time intervals. It is then possible to step through the finite-difference equations both spatially and temporally until a complete solution is obtained (Harleman, 1966).

Finite-difference schemes may be explicit or implicit. In the former, the solution is approached one point at a time. Each known value at a grid point at the time step $t + 1$ can be expressed as a value of other known values at some previous time step t. The implicit scheme requires the solution of a set of simultaneous algebraic equations since each unknown at time step $t + 1$ will depend on other values at the same time step. The application of the finite-difference method to the Bay of Fundy system is described by Greenberg (1976).

Finite-Element Method The area under consideration is broken up into a series of cells of arbitrary shape, a procedure that permits a concentration of cells in areas of prime importance with the common boundary of adjacent cells known as links. Water movement through the links is governed by the momentum equation and is related to the water levels in the cells. These levels, in turn, are determined by application of the continuity equation to each cell.

The specific area to be studied and the phenomena to be investigated will be the primary determinants of the most suitable approach, that is, whether the flooding and drying of mud flats is an important consideration, what size and type of grid system would best cover the areas of prime interest, how far the model should be extended to have independent boundary conditions, and whether or not wind stresses, atmospheric pressure, or Coriolis effects have to be simulated.

7.3.2 The Finite Grid Element

With the type of model decided, the area under investigation must be schematized. For two-dimensional models, therefore, a grid is developed for the areas of interest with extensions to the boundaries (Fig. 7.2). At this point, a major decision is required since accuracy and computational costs are directly related to the size of the grid. The finite-difference model may require different sized grids over different areas of the estuary, namely a fine grid might be devised to cover a specific area of interest and a coarse grid used to extend the simulation to the boundaries. Figure 7.3 illustrates the four grid sizes that were used to model the Bay of Fundy–Gulf of Maine system, with the smaller grid sizes being used for those areas of particular interest. A coarse mesh with $\Delta S = 21.14$ km was used at the mouth of the Bay of Fundy and varying with latitude covered the Gulf of Maine out to the Continental Shelf. For the main part of the Bay of Fundy, a medium mesh size, $\Delta S = 7.047$ km was used with

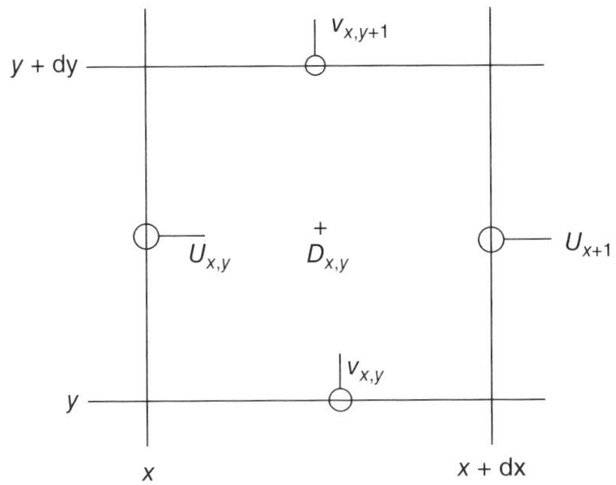

Figure 7.2 Definition of finite grid element.

a fine mesh of $\Delta S = 2.349$ km for Chignecto Bay and the area around Cape Chignecto. A still finer mesh of $\Delta S = 1.66$ km was found to be necessary for Minas Channel and Minas Basin. Methods of joining the grids are given by Greenberg (1976). At this point, the modeler is faced with a dilemma since accuracy is directly related to the size of the grid or cells. However, once this decision has been made and the grid laid out, depth information at each point must be transferred to the computer for use in the numerical solution. If the flooding and drying of tidal flats is important, the computation scheme chosen should be able to incorporate this type of dynamic boundary in the simulation. Depth values for grids extending over these flats have to be determined with great accuracy so that their inclusion in the computation will occur at the proper times in the tidal cycle.

7.3.3 Calibration

The numerical model of an estuary is calibrated by adjusting the bottom stress so as to reproduce a known set of conditions, generally water elevations and phases measured in the prototype for a known set of boundary values. Velocity records may also aid in the calibration. The calibration procedure is analogous to that for a hydraulic model where roughness strips are introduced in the model to reproduce the flow regime being modeled.

The use of fairly long data periods eases the calibration process, particularly for hydraulic models. For numerical models there is no difficulty in simulating a 28-day prototype period. For hydraulic models, the development of the tidal synthesizer, capable of generating very large repetitive periods, can form part of the direct digital control system of the model. Therefore, by calibrating over long periods, minor unnecessary adjustments, due to local effects recorded by the prototype gauges on the day selected for calibration, can be avoided.

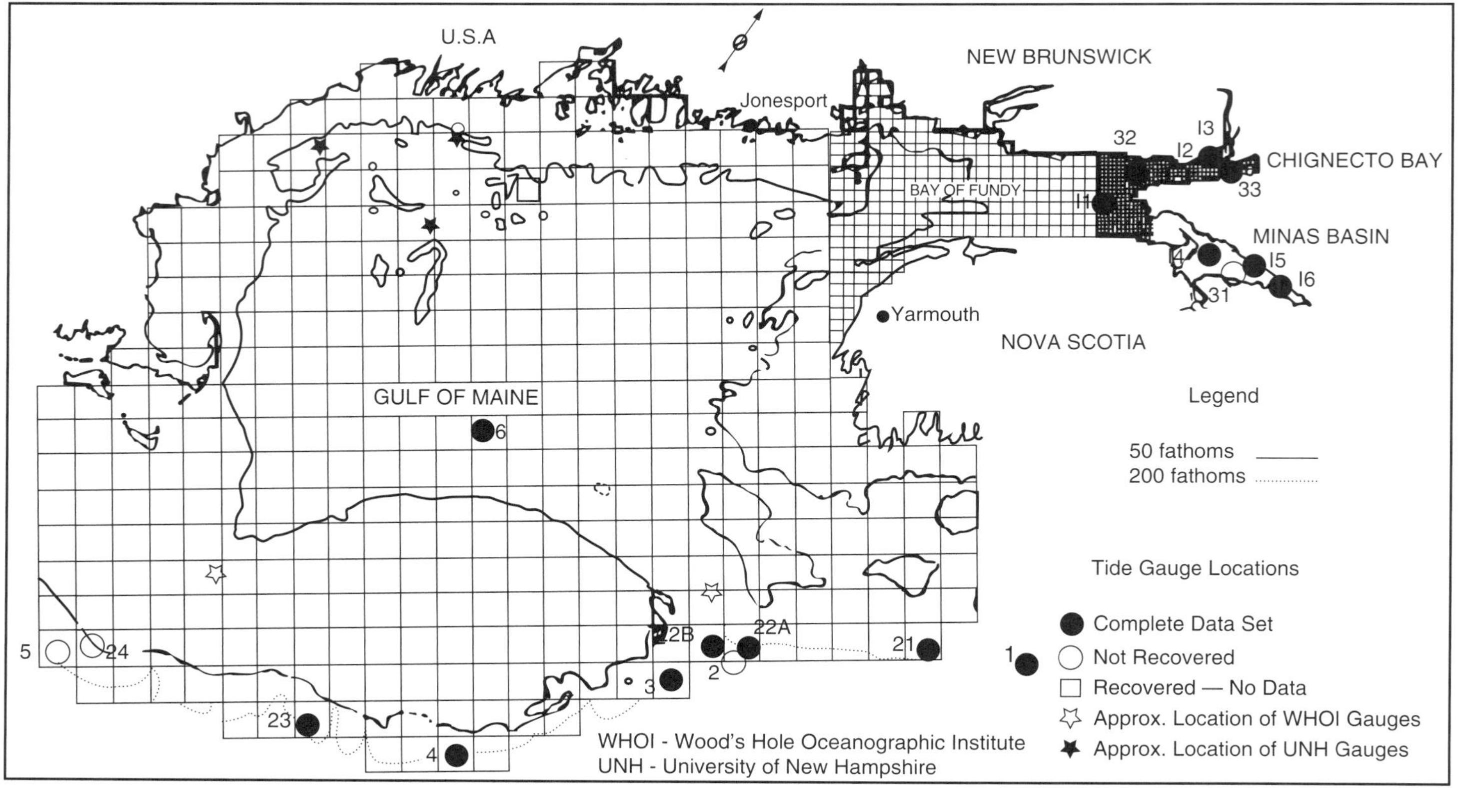

Figure 7.3 Extent of Bay of Fundy–Gulf of Maine mathematical model for tidal regime studies.

7.4 HYBRID MODELS

Hybrid models retain the advantages of both the physical and mathematical simulations at a substantially lower cost (Funke, 1981). For example, a hydraulic model to simulate the effects of a tidal-electric plant at the mouth of Cumberland Basin in the upper reaches of the Bay of Fundy would have to include not only that basin, but all of the Bay of Fundy and Gulf of Maine out to the Continental Shelf. Figure 7.3 shows the area that it was necessary to model mathematically in order to include the whole of the system that is close to resonance with the semidiurnal tide. It was assumed and confirmed, that disturbance of the tidal regime by a barrier or the operation of a tidal-electric plant at the mouths of any one of the upper bays of Fundy would be dissipated or be negligible by the time it reached the seaward limits selected for the model. A hydraulic model of such an area at scales that would yield accurate quantitative results, particularly for any nonlinear conditions, would be impractical.

The alternative approach would be to construct a hydraulic model of the particular area of the barrage location at such a scale factor that distortions would be small and two-dimensional flow patterns would be meaningful, and to interface dynamically this physical model with a numerical model that would extend as far afield as necessary to ensure effective dissipation of reflections at its outer boundary. If the whole of the upper estuary above the barrage could not be included at the hydraulic model scale required for accurate modeling of the barrage effects, closure velocities, simulation of the operation of the powerhouse and sluiceways, and so forth, then a numerical model of the upper portion of the tidal basin could also be interfaced with the hydraulic model. As implied, information is exchanged between these two models almost instantaneously so that both run together as one large model of the whole area. The hydraulic model would handle all those nonlinear problems for which it was designed and the numerical model would look after the dissipation of reflections that it can manage quite well, namely the hybrid mode which eliminates the "boundary" problem (Funke, 1981).

7.5 MODELING OF BARRIER EFFECTS

As pointed out in Chapter 2, tides are responses to the net generating force resulting from variations in the gravitational attraction of the moon and the sun and vary, not only with the intensity of this net force from time to time, but also from place to place due to physiographic conditions. Thus, the range of tides in the headwaters of a relatively long estuary would be affected both by the alteration of its physiography implied by the construction of a barrage and by the extraction of energy from the tidal waters.

To investigate such changes, the numerical model, calibrated to simulate the natural tidal regime in the estuary, can be run with the introduction of modifications to simulate the changes in estuary configuration and associated flow changes arising from the construction and operation of a tidal-electric plant in the estuary.

The effects of a permeable barrier, representing the action of a tidal-electric plant are important and should be determined as accurately as possible. A firm estimate of the cost of a plant's output must be based upon the best possible understanding of the tidal ranges likely to prevail as a result of its operation. Nevertheless, the effects of a tidal power development when the plant is not operating, that is, when it acts as an impermeable barrier, are also important to define its seaward effects.

One feasible approach to simulating a development and its operation is described in the reports on the Canadian federal/provincial reassessment study of the Bay of Fundy tidal power potential (Bay of Fundy Tidal Power Review Board and Management Committee, 1977), and is based on the equation of flow presented by Heaps (1972) as follows (with reference to the notation of Fig. 7.4).

For seaward discharge through turbines,

$$Q_n = K(\Delta\xi_n)^{1/2} \qquad \Delta\xi > h_b \tag{7.4}$$

With no flow,

$$Q_n = 0 \qquad 0 < \Delta\xi_n < h_b \tag{7.5}$$

For reservoir filling

$$Q_n = K'(|\Delta\xi_n|)^{1/2} \qquad \Delta\xi < 0 \tag{7.6}$$

where $\Delta\xi_n$ = head of water from the level in the reservoir immediately behind the barrier to the level on the seaward side of the barrier at section n

K_n = constant of flow at barrier section n relating the flow through the turbines (during power generation) to $\Delta\xi_n$

K'_n = constant of flow at barrier section n relating the flow through the sluices and turbines (on reservoir filling) to $\sqrt{|\Delta\xi_n|}$

Q_n = flow (positive seawards) across grid section n of a barrier site

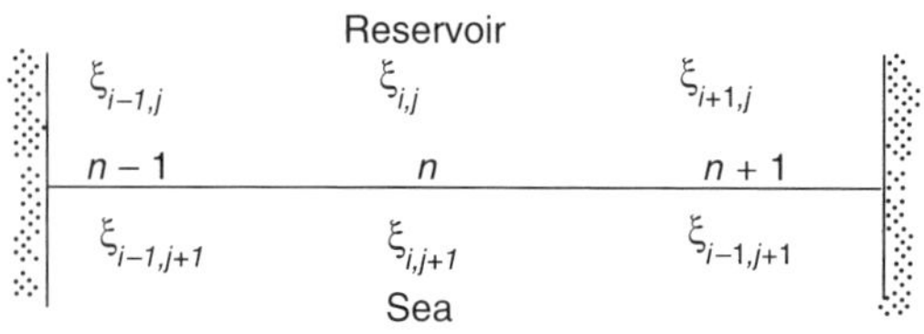

Figure 7.4 Notation for determining effects of tidal barrier.

h_b = minimum head difference necessary for power generation
$\Delta\xi$ = head of water from the mean level on the reservoir side of the barrier to the mean level on the seaward side

The effects of a tidal power development on the physical oceanography of the Bay of Fundy–Gulf of Maine system are also described by Greenberg (1977).

7.6 MATHEMATICAL MODEL FOR CLOSURE ACTIVITIES

The construction of a barrage across an estuary or other area of tidal water constitutes the *closure* phase. Once construction activities start to block the waterway, the tidal currents through the remaining area increase to maintain the tidal flows to and from the basin that is being isolated. The increasing tidal velocity requires an increasing head difference across the barrage to force the water through it. This causes the basin tide to lag behind the open sea tide and ultimately to reduce in range when the basin can no longer be completely filled or emptied before the sea tide turns. The increasing head differences and tidal velocities during the construction sequence of a tidal barrage pose potentially serious problems. At all stages of closure of the estuary, and more particularly during the final closure stage whether this is accomplished by *end filling*, by a *blind* caisson, by a temporary sluice caisson, or by raising the crest of a closure dike, the elements of the barrage must be able to withstand the hydraulic pressures and velocities arising from these head differences. This aspect of construction of a tidal power development is discussed more fully in Chapter 8. Flat-surface mathematical models that assume that the water level in the basin being isolated and the tide level on the seaward side are known, are easily developed and prove very useful in prefeasibility studies to determine alternative construction sequences, order of placing caissons, and the like. The input parameters would include tidal range, closure width, barrage height and basin storage area, and effective number of sluices. The model would determine an average velocity for a given opening and constant barrage heights; the flow would be simulated over a full tidal cycle until an equilibrium condition is obtained.

7.7 UTILITY SYSTEM PLANNING AND SIMULATION

The economic analysis of a tidal-electric development involves the studies of costs and benefits. Mathematical models are available to allow the impact of storage devices and transmission capacity between market areas to be studied and optimized. Mathematical models are also available to determine the amount of raw tidal energy that can be absorbed by a power system. A discussion of examples of such models is given in Chapter 12.

CHAPTER 8

CIVIL WORKS FOR TIDAL POWER DEVELOPMENT

8.1 INTRODUCTION

The major elements of a tidal power development include the barrage, a tidal basin, and a transmission link to the power system grid. The primary purpose of the barrage is to close off an estuary from the sea so as to create a head by controlling the water levels of the basin thus formed, at elevations lower than high tide or higher than low tide. The concepts of development schemes are outlined in Chapter 5.

In many respects, a tidal barrage is a much simpler and cheaper structure to install than a river dam of similar proportions. The main physical difference between them is that the barrage retains water on both sides, hence the static forces imposed upon it are smaller. Moreover, the leakage of water beneath or through a barrage is not as important as for a river hydroelectric dam, so long as such leakage would not lead to significant energy losses nor risk the integrity of the structure.

The barrage itself consists of a powerhouse, a sluiceway section, and dike connections (access and closure) to the shore to form the controlled tidal basin. Once the approximate barrage line has been selected based on considerations of energy output, cost and environmental impacts, the exact line then needs to be selected to take account of the contours and geology of the estuary bed, the size of the powerhouse and sluiceway elements and local environmental factors.

Elements of Tidal-Electric Engineering. By Robert H. Clark

8.2 DRY VERSUS WET CONSTRUCTION

Structures built in rivers are normally erected within cofferdams, which create dewatering so that the powerhouse and sluiceway structures can be erected "in the dry." This was the technique, illustrated in Fig. 8.1, used in constructing the tidal-electric plant in La Rance Estuary, France (Mauboussin, 1972). The site was relatively clear of sediments and there was little commericial navigation with which to contend. Nevertheless, the cofferdams accounted for about 30 percent of the cost of the project's civil works (Severn Barrage Committee, 1981a).

An alternative approach requires the construction of the barrage to be carried out "in the wet." Advances in construction technology associated with marine structures have conclusively demonstrated the feasibility of using concrete caissons that have been fabricated offsite and then floated into place. Operations with large, floated-in structures have been carried out successfully in constructing such projects as the Arromanches Harbour, Normandy, France (Jellet, 1948); the Delta Plan Works in the Netherlands (Fig. 8.5); the Deas Island (Vancouver, Canada), Chesapeake Bay (United States), and La Fontaine (Montreal, Canada) tunnels; the siting of oil production platforms

Figure 8.1 View of La Rance construction site (completed navigation lock in foreground and sluiceway in background). (Courtesy Electricité de France.)

in the North Sea during the 1970s, and in the Atlantic Ocean, 315 km east of Newfoundland, in 1997 (Fig. 8.2); the construction of the Storm Surge Barrier (Fig. 8.3) during the 1980s to protect St. Petersburg, Russia, from high-intensity water surges that occur in the eastern part of the Gulf of Finland and the Neva Delta; and the construction of large, run-of-river hydroelectric plants.

In the late 1960s, an experimental tidal power plant was constructed in a small bay off the Barents Sea, Russia (refer to Chapter 15) by a simple two-unit caisson (Fig. 8.4) that was fabricated near Murmansk, towed some 90 km to the mouth of Kislaya Guba, and sunk onto a prepared foundation at a site about 50 m wide where the tides range from 1.3 to 3.9 m. According to Bernstein (1972) the powerhouse module was placed on the prepared foundation within a few millimeters of the correct vertical position and only a few centimeters from the proposed horizontal position.

Floated-in caisson sluiceway units have also been successfully placed to form part of the Scheldt Estuary Control Works, one element of the overall

Figure 8.2 The Hibernia concrete caisson platform, at the deep-water site, Bull Arm, Newfoundland, with all nine tugboats in place and ready to tow, May 22, 1997. (Courtesy Hibernia Management and Development Company Ltd.)

Figure 8.3 Towing one of the concrete sluice-gate caissons to the site of the St. Petersburg Storm Surge Barrier. (From Bernstein, 1995.)

Figure 8.4 Floating in and aligning the Kislaya Guba tidal power station caisson at the prepared site. (From Bernstein, 1991.)

Figure 8.5 Closing the gap in the dam at the Veerse Gat, part of the Scheldt Estuary Control Works. (From Clark, 1972.)

Delta Plan Works in the Netherlands. Figure 8.5 presents an aerial view of the last concrete caisson for the dam at Veerse Gat being towed into position for sinking.

The float-in technique has also been used in the past two decades for constructing low-head hydroelectric plants and has been suggested for installing powerhouses at navigation dams in the United States that are currently without power facilities (Makela, 1983). A recent application of this technique consisted of the construction, in New Orleans, Louisiana, of a prefabricated steel caisson containing eight 24-MW generators driven by geared-type bulb turbines. The caisson, comprising the preassembled plant structure measuring 44 m in width, 139 m long, and 38 m high and weighing more than 21,000 tonnes, was towed up the Mississippi River (Fig. 8.6) to its final destination near Vidalia, Georgia, about 320 km upstream from New Orleans and sunk onto prepared foundations (Hadley and Lindestrom, 1990).

Using this floated-in technology rather than construction in situ that is, in the dry, the elements can be manufactured away from the site, floated into position, and then sunk onto prepared foundations. The technique offers a viable alternative to the necessity of cofferdaming a site and, with facilities readily available for construction and towing, a properly coordinated program

Figure 8.6 Towing the *Murray* prefabricated-steel hydroelectric power station up the Mississippi River. (From Makela, 1983.)

would offer considerable economies. In the reassessment studies of the Bay of Fundy (Canada) tidal power potential, the comparative economies of these alternative construction concepts for the three selected sites were assessed. Alternative cofferdam layouts were evaluated to achieve the best balance between costs for the cofferdams and the rock excavation. It was concluded for those selected sites that construction in the dry behind temporary cofferdams would cost a minimum of one-third more than the use of floated-in caissons and, at one of the sites, more than twice as much (Bay of Fundy Tidal Power Review Board and Management Committee, 1977).

In summary, there are three major advantages provided by this technology: (1) Economies can be achieved through acceleration of the program of installation, yielding a significant reduction in the schedule of construction; (2) substantial construction work can be carried out in locations remote from the site so as to reduce significantly the large concentration of resources accompanying in situ construction; and (3) environmental impacts are minimized, particularly with respect to sediment movements during the construction period (Warnock and Clark, 1992).

8.3 DESIGN PARAMETERS

The requirement to float a caisson unit into place on the barrage line will require a greater width/depth ratio than would normally be the case for in situ structures, but this will also be effective in achieving lateral stability. Moreover, the floating stability of the unit must be carefully assessed and equipment installed for controlling the immersion of the structure on location. However, before consideration can be given to the design of caissons or of the dikes or embankments, data must be available to determine the integrity of foundation conditions and to derive the hydraulic design parameters.

8.3.1 Foundation Considerations

Bedrock foundation conditions can only be adequately evaluated on the basis of geological surveys, geological mapping, and core drilling in the vicinities of the selected sites with laboratory testing of the samples (Walia et al., 1979). For reconnaissance-type assessments of tidal power possibilities, foundation conditions may be inferred from existing data but, in most cases, such data will be nonexistent or so sparse as to make deductions from them very tenuous. Seismic surveys, which are quick and relatively cheap to carry out, can now provide a fairly dependable source of data for reconnaissance and preliminary investigations.

It would be unlikely that, at a chosen site, a competent bedrock foundation would be found at an appropriate elevation, that is, without excavation. On the other hand, where rock is not available at a convenient elevation, foundations on overburden would be acceptable with the designs adapted to suit the conditions. Based on the available data, augmented as appropriate, reasonable estimates or assumptions are required for the following aspects of foundations: (1) bearing capacity, (2) settlement, (3) sliding and uplift, (4) piping, (5) percolation and (6) scouring.

8.3.2 Hydraulic Considerations

The primary areas of hydraulic concern for design and construction are the maximum and minimum water levels and tidal currents to be expected at barrage sites.

Maximum Water Levels Crest elevations for all structural elements of a tidal power development depend upon assessments of the maximum water levels to be experienced at the barrage. Such assessments must take into account maximum tidal levels, wind setup, storm surge, wave runup, and any long-term variation in mean water levels that may be anticipated within the estuary. Maximum tide levels to be expected at a permanent tide gauge location are usually published annually as *higher high-water levels for large tides* (HHWLT). Values for the estuary or site locations may be interpolated from such published data for the region. Such tides can be expected approximately once monthly.

The wave climate at each site location must be determined. Deep water and wave height predictions (U.S. Army Coastal Engineering Research Center, 1973) can be made from data from wave rider buoys and long-term wind records at land-based stations in the vicinity and may be determined from the wind spectrum and fetch lengths computed by Saville's (1958) method. A software system is now available that incorporates different algorithms for the analysis of combined environmental extremes (Nessim et al., 1991).

The wind data from land-based stations, for application to the overwater arc at the site, should be increased by as much as 20 percent, or more, to obtain the overwater velocities. The wind spectrum or *wind rose* can be prepared to illustrate the wind climate graphically. The wave climate or the behavioral height characteristics of waves under the influence of wind is determined by a refraction analysis for each barrier location. The significant wave height is computed as the mean of the largest one-third of the waves for each storm condition evaluated. The maximum individual wave may be about twice the significant wave height.

To provide adequate safety in the design of barrage elements, it should be unnecessary to select crest elevations for all structures such that overtopping would not occur under any combination of adverse conditions; indeed, it would not be an appropriate engineering approach. During stormy weather, access along the barrage should be possible for the personnel in charge of the plant. Thus, the crest of the embankment would have to be chosen to avoid a "solid" wall of water running up the embankment face and reaching the road. The following three levels of security concerning overtopping of the dikes provide a basis for selection of the appropriate criteria:

1. *Severe* The crest elevation would permit waves greater than the significant wave to overtop the crest. For this condition, allowing for runup, the crest could be set at an elevation 1.2 times the significant wave height above maximum still-water levels.
2. *Moderate* The crest elevation would be selected so that only 5 percent of the waves reach the crest. For this condition, the crest could be set at an elevation 1.5 times the significant wave height above maximum still-water level.
3. *Light* The crest elevation would allow only 1 percent of the waves generated by the design wind to reach the crest. For this condition, the crest could be set at an elevation 2.2 times the significant wave height above maximum still-water level.

Runup of waves on a vertical face is assumed to be about 1.75 times the significant wave height, although the splash zone would extend well above this level. Runup depends on the structure's shape and roughness, the water depth at the toe of the structure, as well as the slope of the seabed. A

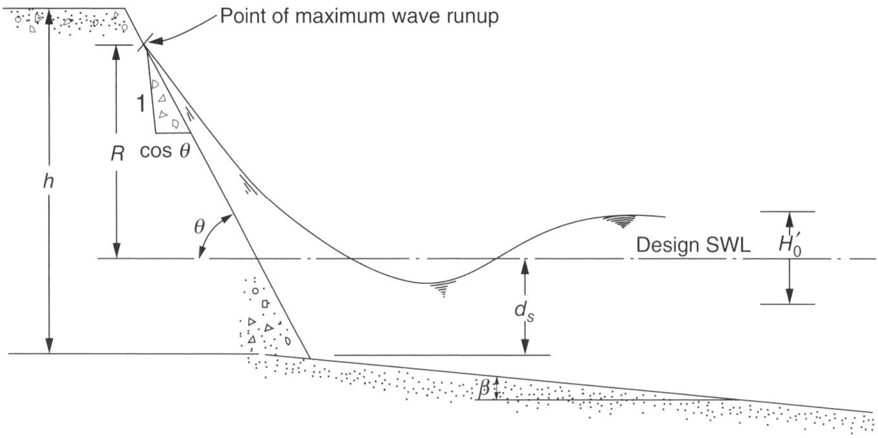

Figure 8.7 Definition sketch of wave runup.

definition sketch of wave runup is shown in Fig. 8.7. Referring to the sketch, if the wave height, $H_0' = 10$ m with an embankment slope of 1 on 2 and a still-water depth, $d_s = 20$ m, then R would be about 22 m for a smooth surface, 10 m for rip-rap, and 11 m for dolos (refer to Fig. 8.16).

If not already available, good overwater wind speed data should be obtained to support detailed designs. A variation of the order of 1 m in the determination of the design wave height and the overtopping criteria is possible, and could affect dike costs, including armor requirements, by as much as 10 percent. On the other hand, this might represent no more than about 3 percent of overall development costs.

Minimum Water Levels The minimum water levels at barrage sites establish the elevations of protective armor layers on dikes and are also needed for selecting turbine settings. Such minimum levels may be computed from tabulated tidal data and estimates of wind setdown and wave drawdown. Minimum tidal levels would be the *lower, low-water level, low tide*, or LLWLT. The same wind setup and significant wave evaluations used for selecting structural crest elevations can be used in the opposite sense for assessing minimum water levels. For single effect, ebb-tide generation, avoidance of turbine cavitation is a critical consideration at the point in the operating cycle when the seaside water level is at its lowest. The minimum seaside water level would be the minimum tidal level less a wind setdown, with an appropriate exceedance frequency determined from the wind spectrum analysis. The turbine manufacturer will specify a minimum setting for the turbines below the lowest water level that would produce no cavitation. However, the severity of the cavitation condition allowed depends upon the actual conditions of operation of the turbines at the time of criticality.

Tidal Currents Tidal current data form the basis of evaluations of several construction procedures including the floating into place of caissons and the placing and final closure of dike sections. In the high tidal ranges at prospective tidal power sites, very careful scheduling of the various stages of construction is necessary in order to keep the civil engineering costs as low as possible. A mathematical model can be a very effective tool in predicting water velocities as the tidal basin is closed. Local velocities and current patterns may have a significant bearing on construction methods and procedures, particularly for the placement of caisson units.

8.4 CAISSON DESIGN

The caisson must, of course, have stability not only while it is being towed into position but also after it has been sunk onto the prepared foundation. It must then withstand the differential hydrostatic and wave forces to which it will be subjected. The latter stability can be achieved by penetration into the foundation of "skirts," integral with the caisson, or similar anchors, to depths limited to 5 m or so. Dredging to suitable subsoils would be carried out prior to installation of the structure. Derrington (1979) has pointed out that, based on experience with the gravity structures placed on the North Sea floor, a skirt penetration of 5 m should be able to be achieved without difficulty so that design wave and other environmental loading on the order of 5000 kN/m of length can be resisted with adequate safety. Careful design of the floating stability of the unit must be made, and equipment, such as valves and cocks, installed for controlling the immersion of the structure on location.

The construction activities or procedural steps for construction of the barrage would generally include:

- Dredging and removal, as required, of overburden and soft soils unsuitable as foundation materials for dikes, and placement of gravel mattresses under the powerhouse and sluiceway elements and of erosion-resistant material upstream and downstream of these elements
- Building of dry docks and construction yards
- Placing of sand and gravel mattresses for the elements
- Placing crushed stone scour protection to protect the mattress
- Floating-in and setting the powerhouse, sluiceway, and blind caisson elements
- Installing turbogenerators and mechanical/electrical equipment in the caisson units
- Installing precast concrete closure panels between, and at the ends of, the caisson walls and backfilling with gravel
- Grouting up the scour protection areas and under the base of the elements

- Ballasting the pockets in the powerhouse and sluiceway elements with sand
- Placing rip-rap over the mattress upstream and downstream of the elements as protection against current and wave action
- Completing construction of the dikes using a vertical closure method.

The order, and even the appropriateness, of some of these steps would depend upon the specific site and the comparative economics and scheduling feasibility of carrying out some of the activities separately or in combination. For example, the turbogenerators could be installed before the powerhouse caissons are floated to the site, or blind and open caissons or other techniques may be used to substantially shorten the embankment and final closure sections. From analyses of manoeverability and control requirements during placement of caissons, water velocities greater than 1 m/s would be regarded as unacceptable. A "blind" caisson or caissons could be placed between any dikes or embankments and a powerhouse or sluiceway caisson to minimize interference of high-velocity flows through the discharge passageways with the material at the toe of the dikes and also to minimize the possibility of material entering the passageways.

The major considerations for the structural integrity of the caissons are their stability against sliding, overturning, and seepage. The minimum length of a caisson, that is, from sea side to basin side, would be governed by the more restrictive of: (1) antiseepage reguirements or (2) antisliding or overturning. Figure 8.8 illustrates a preliminary stability assessment to develop an estimate of the lengths of powerhouse and sluiceway caissons as a basis for an evaluation of the tidal power potential and construction characteristics of an estuary. The minimum length of the structure would be governed by seepage. On the other hand, the assumed foundation conditions might lead to the conclusion that the maximum length would be governed by the factor of safety against sliding and overturning and the allowable bearing capacity of the mattress and overburden. It would be necessary to ensure that all required electrical and mechanical equiment, in addition to water passages of adequate dimensions, could easily be accommodated within the length calculated to provide stability against sliding, overturning, and seepage.

Site investigations are required to obtain foundation data on which to base a design to confirm engineering feasibility. Minimum factors of safety under various critical or worst loading conditions are suggested in Fig. 8.8 for prefeasibility studies. Other factors such as ice, wind, waves, saltwater attack, temperature extremes, and expected seismic conditions must also be considered at the appropriate time.

Sliding failure of the caisson units should be checked both at the level of the prepared sand and gravel mattress foundation and at the interface of the prepared foundation with the accepted overburden. However, such overburden could contain silt seams or lenses having low permeabilities so that for undrained conditions a low internal friction angle could occur. A

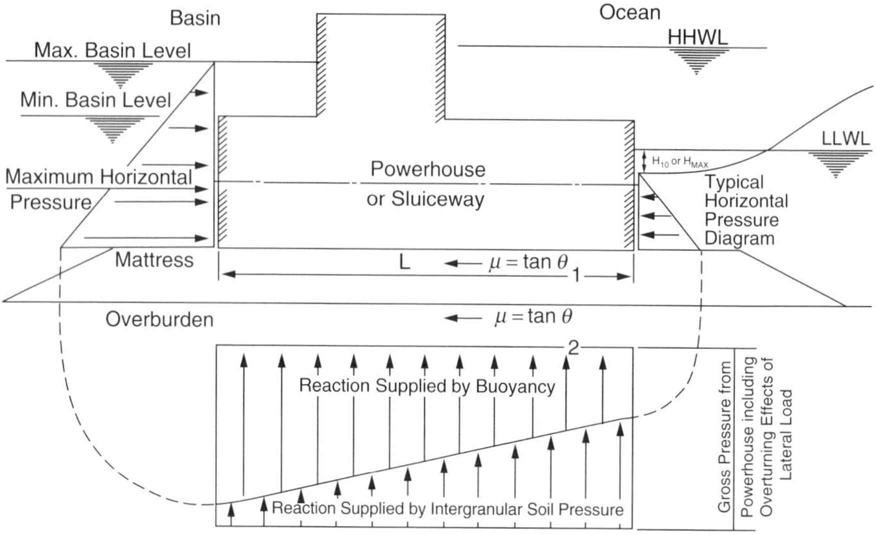

Notes

H_S significant wave height

H_{10} 1.27 H_S

H_{max} 1.94 H_S

Critical Loading Cases Considered

Loading Case	Differential Head, H	Factor of Safety
Horizontal	(Max basin - LLWL) + H_{10}	1.5
Horizontal	(Max basin - LLWL) + H_{max}	1.25
Vertical	see diagram	1.6
Seepage	(Max basin - LLWL)	2.0

Figure 8.8 Caisson stability forces.

design value of 30° might be assumed until confirmed by site investigations. The sand and gravel mattress is designed to provide safety against piping failure; it would serve as an inverted filter and must, therefore, be filter-graded accordingly.

As noted previously, the caisson would be fitted with a skirt to mobilize sliding resistance, and the underlayer would be grouted. Founding caissons on the seabed subsequent to dredging could be accomplished by means of: (1) a small number of large-diameter, bored stub piles or (2) a large number of small-diameter, interlinked hydraulic jacks built into each caisson. Both methods would initially support the caisson at a predetermined level and allow the space beneath the caisson to be filled with concrete. After the concrete had set, the placing of the permanent ballast in the caisson would be completed. Derrington (1979) concluded that, as a result of advances in the design, construction, and placement techniques of concrete sea structures, concrete caissons could be placed to an accuracy of 20 to 25 cm in the relatively shallow and sheltered waters of a river estuary. However, subsequent to the 1970s, placement techniques have been further enhanced, particularly during the completion of the Delta Works as a result of the application of laser positioning of the ships and computer technology in ship propulsion, so that caissons can now be placed within a tolerance of about 5 cm.

The positioning of a caisson will generally be carried out during the time of slack water, at high or low tide. High water would be a convenient time to bring a deep-draught caisson to its prepared foundation. A preliminary ballast adjustment, allowing only the necessary bottom clearance while being brought into position, would minimize the amount of vertical movement after the caisson is correctly located above its foundation. Extensive investigations are required to determine the patterns of water movement about the partially completed works so that the necessary operations can be planned and scheduled (Gwynn et al., 1972).

After each caisson had been lowered into position and leveled, the space beneath the caisson would be filled with concrete to prevent water flowing through the gap and to spread the load of the caisson over the foundation. Before filling could start, however, it should be verified that the skirt is embedded in the foundation material so that the undercaisson concrete would be retained. Bolsters, inflated with grout, could be used because they would conform to reasonable undulations in the foundation. The bolster would be attached to the underside of the caisson in the construction yard. As more caissons would be placed and the differential heads across the barrage increased, a second, or backup, seal could be added to subsequent caissons. The progress of concreting under the caisson could be monitored by adding a radioactive tracer to the mix to track it with sensors—a technique used successfully to track the grout linking the legs of the North Sea structures to their piles. The sensors could be linked by microprocessors to control valves so as to admit concrete through different pipes as the concrete front advanced.

In choosing the caisson-type or floated-in construction for the main portion of the barrage—because it requires the lesser construction time and would generally offer the more economical solution to the alternative of construction behind cofferdams—the uncertainties and risks associated with this technique should be fully comprehended. The preparation of the foundation, towing of the caissons, and their placement onto the prepared foundation, although seemingly simple, will require special equipment and procedures for each site according to the tidal characteristics and sediment movement encountered. Very close monitoring of the foundation excavations, together with the results of core drilling, will permit designs that would virtually eliminate risk by permitting adjustments to the preparation of the foundations to match local conditions. Engineering management (refer to Chapter 4) must impose stringent quality control during construction as well as an exhaustive program of monitoring after caisson placement. Prototype testing of a selected procedure is recommended since this is probably the most difficult aspect of the construction of a large-scale tidal power development in estuaries with high tidal ranges, such as the Severn Estuary and Bay of Fundy.

8.4.1 Powerhouse Caissons

For preliminary cost estimates the turbine caisson may be designed to house two, three, or four turbogenerators, the choice depending upon the diameter of turbine selected, overall cost, construction time, and stability during towing. The caissons could be fitted with the turbines and generators together with the associated control equipment, switch gear, transformers, and cabling. The caissons would then be towed to their final position where each would be ballasted onto the previously prepared seabed, and the gap between the underside of the caisson and the seabed, sealed. A typical arrangement for a proposed two-unit powerhouse caisson for Bay of Fundy sites is shown in Fig. 8.9, and the sectional elevation of a proposed four-unit powerhouse

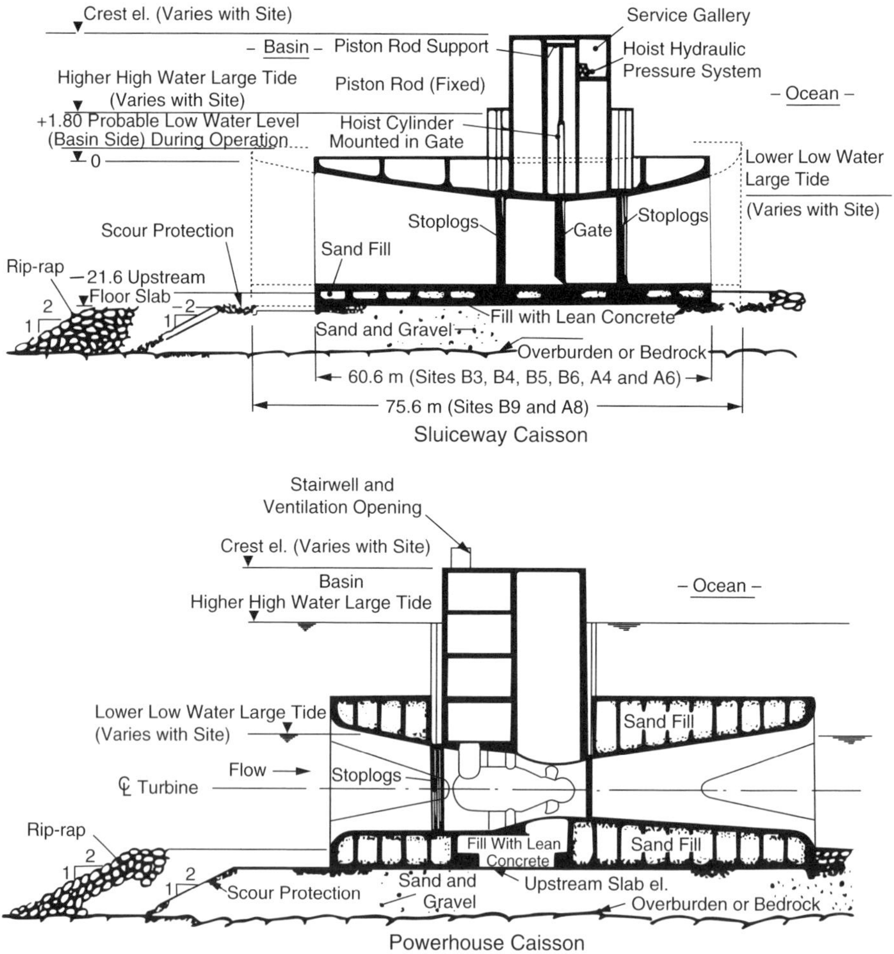

Figure 8.9 Sectional elevation of typical powerhouse and sluiceway caisson units for a Bay of Fundy scheme.

caisson for the Severn Estuary is shown in Fig. 8.10. It should be kept in mind when determining draft tube submergence that oscillatory effects may be induced in the draft tube, as a result of loss of suction head, even when sea level is slightly above the top of the tube. Waves could also trigger these effects, resulting in a substantial reduction in power output.

8.4.2 Sluiceway Caissons

The sluices are incorporated into the barrage of a tidal-electric development to control water levels. In single-effect operation, the sluices are used to fill the tidal basin on the flood tide and are then closed so that the turbines have a positive generating head during the ensuing ebb tide. Double-effect schemes may use the sluices to pass flows in either direction to increase the head differential for generation or reduce it during times of pumping.

Unlike conventional hydroelectric spillway gates, the sluices of a single-effect, tidal-electric development, must be opened and closed in accordance with the diurnal tidal regime, or 705 times annually. Thus, the gates must be capable of rapid, frequent operation and be as free as possible from mainte-

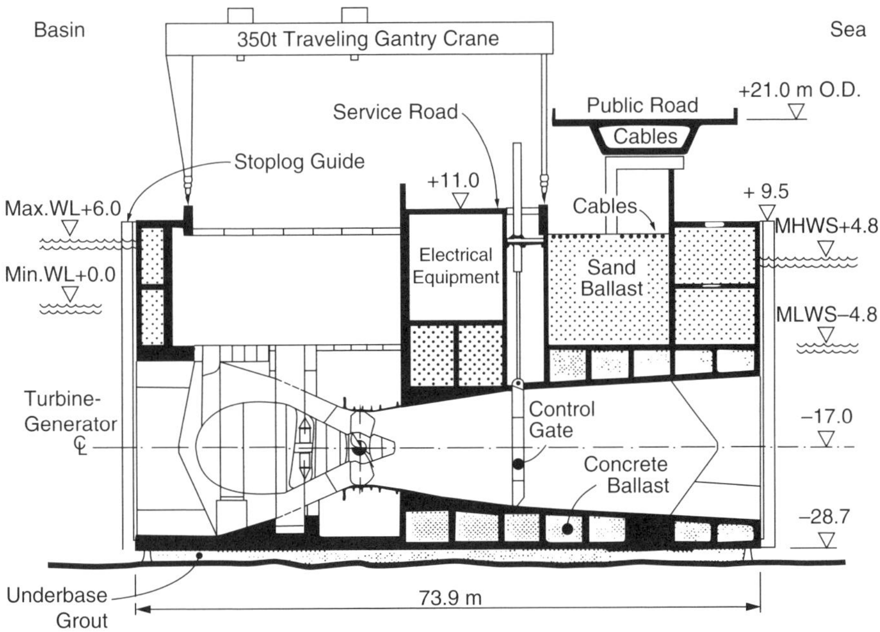

Figure 8.10 Sectional elevation of proposed typical four-unit powerhouse for Severn Estuary scheme. (From Severn Tidal Power Group, 1989.)

nance and operating problems. It is also essential that the gate setting ensures adequate submergence at all times to avoid damage or operational difficulty through wave impact and corrosion of the gate mechanisms and, if applicable, ice jamming and freezing.

The major design criteria for sluiceway gates are:

- Capable of being opened or closed against small differential heads so that the operation of the sluices can be spread over a time period to suit controlled startup or shutdown of electrical machines (15 to 30 min).
- Capable of being opened and shut in less than about 15 min.
- Capable of being operated against larger differential heads than would occur under normal operation so that they can be opened later than normal should power failure occur at the slack water period.
- In situ maintenance should be a minimum: Therefore, gates should be capable of being easily removed and should be robust and protected against corrosion and abrasion.
- Each passageway should be capable of being dewatered for inspection and maintenance.
- Capable of easy release should they become jammed (either by malfunction or floating debris).
- Clearance of large floating debris or ice floes obstructing gate openings should be easily accomplished.
- Sealing against water leakage is only necessary to prevent gate vibration and consequential corrosion and erosion.
- When open, gates should not unduly impede flow of water through the sluice openings.
- Power required for gate operation should be kept to a minimum.
- Capable of functioning in conditions of ice formation.
- Able to work effectively in the event the caisson is not perfectly level.

The types of gates most suitable for tidal-electric operation are: radial gates and vertical-lift gates. Flap gates, although requiring no operating mechanisms, have a number of serious disadvantages for tidal power operation and are not considered suitable for large developments (refer to Severn Barrage Committee, 1981a).

A radial gate caisson proposed by the Severn Tidal Power Group (1989) is shown by Fig. 8.11. The gate incorporates buoyancy chambers and is expected to give significantly lower maintenance and operational costs than other types.

Vertical-lift gates could be operated to suit overall barrage requirements, including flood control. Maintenance of the gates could be carried out above high-water level and, if necessary, they could be removed and replaced through

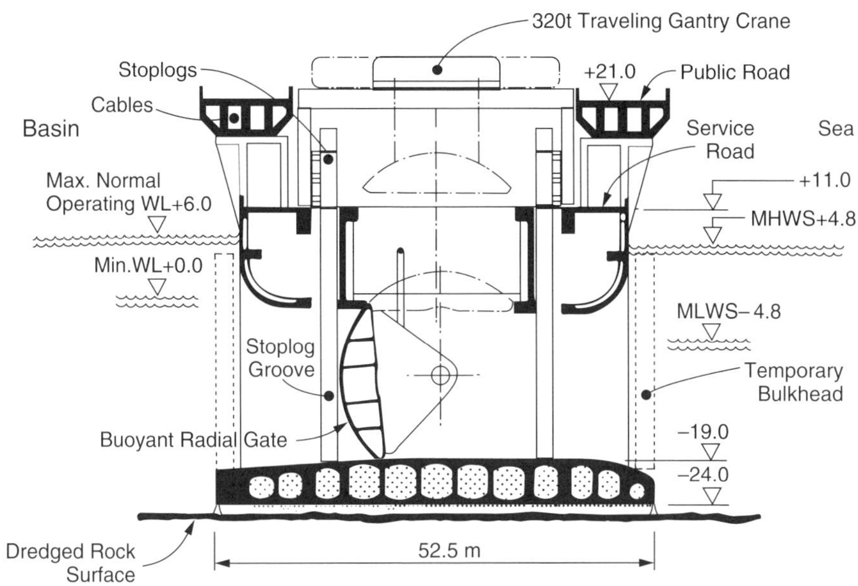

Figure 8.11 Radial-gate sluiceway caisson. (From Severn Tidal Power Group, 1989.)

the top of the caisson by a floating crane. As with sluiceways equipped with radial gates, the sluiceways incorporating vertical-lift gates are equipped with stoplogs to permit dewatering of water passages and access to gate sill and seals (refer to Fig. 8.9).

8.5 DIKES

Following the construction of ship locks (if navigation is a major factor in the estuary) and the towing, positioning, and sinking of the turbine and sluiceway caissons, dikes or embankments complete the barrage. These embankments serve two purposes: access for operation and maintenance of the barrage and a safe route to shore for the transmission cables.

Dikes may represent between 10 and 20 percent of the cost of a barrage that may, in turn, represent about one-half the cost of a development. Therefore, for the purpose of cost estimating in a prefeasibility study, the traditional method of dike construction may be chosen based on proven technology and wide experience. Alternatives may well prove more economical, such as "blind" concrete caissons (simple boxes) subsequently filled with ballast, or open caissons to maintaln low closure velocities with provision for dropping concrete stop logs into position during the final closure activity.

The major factors for which design criteria would have to be specified would include:

- The range of water levels on each side of the dike and the differential heads that could occur during construction and subsequent operation.
- The combination of wave climate, including the maximum height of single waves, and the water levels that would have to be withstood on both sides of the dikes.
- The space required for power cables and necessary protective measures, such as concrete cable ducts.
- The sizes and weights of the largest individual items of plant or machinery requiring access during construction and operation; these might include turbine runners, generators, transformers, and sluice gates, although these could be assembled with the powerhouse or sluiceway caissons before towing to the site.
- The need for wind breaks or other protective measures that would reduce access problems for personnel during extreme weather conditions.
- The strengths, durablity, and other properties of both the embankment foundations and the materials for construction.
- The tidal velocities through, around or over incomplete embankments that could occur during construction.

The foregoing are, of course, site-specific factors and dependent upon the choice of alignment and layout of the barrage. For example, wave height may, in fact, influence the final layout of the barrage. However, for the purposes of a prefeasibility study, judgments concerning these factors must be exercised based upon the available information.

For the main embankment, the major problem to be overcome would be the achievement of control over tidal flows by means of an initial rockfill mound or other method. Techniques that have been used successfully in tidal conditions to build the control structure include:

- Dumping rockfill from bottom dump or side-dump barges and end-tipping rockfill using large dump-trucks and bulldozers.
- Using large cranes, equipped either with grabs or with special skips to place rockfill; the cranes may either be mounted on barges or be land based.
- Using caissons that are subsequently buried in the embankment.
- Using overhead cableways to dump concrete blocks, dolos, or rocks.
- Establishing a steel framework or trestle from which dumping could take place; the core of the dike section would be formed by a wall of precast concrete panels lowered between I-beam framing members to about 5 m below sea level along one side of the trestle.

- Final closure could be achieved by lowering closure gates in each span of the trestle across the full width of the gap during slack tide; the gates would be left in position and the remainder of the dike completed.

The first three techniques are commonly used and all three may be used on a project at appropriate stages in its construction. The information for Table 8.1 has been taken from the report of the Severn Barrage Committee (1981a) to provide a general comparison of alternative conventional methods of building dikes.

8.5.1 Access Dike

This dike would provide operating communications between the powerhouse and sluiceway complex and on-shore abutment and would be constructed initially so as to provide for movement of personnel and equipment during succeeding construction stages. The access dike could initially be extended into a depth of water sufficient to accommodate construction of drydocks in the protected lee area. Following placement of the first caisson elements, the access dike would be completed.

TABLE 8.1 Comparison of Alternative Methods of Building Dikes

Factor	Barge-Dumped Rockfill	End-Tipped Rockfill	Cableway	Caissons
Dependent on state of tide	Yes	No	No	Yes
Dependent on weather	Yes	Some	Some	Yes
Large quantities of rock needed	Yes	Yes	No	No
Navigation involved	Yes	No	Some	Yes
Affects amenity of adjacent coastal areas	No	Yes	Some	No
Extensive concrete work	No	No	Yes with blocks	Yes
Mobilization time	Short	Short	Long	Long
Backup facilities possible	Yes	Yes	No	No
Potential rate of progress	High	Medium	Low	High
Probable ranking in order of increasing cost	1	3	4	2

A typical cross section of an access dike proposed for an ebb generation Bay of Fundy scheme is shown in Fig. 8.12. Typically, it contains a rockfill zone and a transition zone on the basin side to limit permeability, with rip-rap or armor stone on both faces. The basin water level variation during operation is, of course, much smaller than the natural tidal range on the seaward side of the barrier. Under double-effect operation, however, the reversing hydrostatic differential head alternates between the sea and basin sides of the dikes and is approximately equal to the tidal range. The dike design for such operation would reflect these conditions.

Since the access dike would be constructed first, water velocities that would be encountered in the tip area would not be significantly greater than those experienced at the same location under natural tidal regime conditions. It would not, therefore, be necessary to use very large rock materials, and construction by end-dumping or end-tipping would likely be cost effective.

The construction sequence of an access dike for an ebb generation scheme would be generally as follows:

- Soft, loose sediments would be dredged from the foundation area to provide a base of adequate bearing capacity.
- The control rockfill embankment would be constructed by end dumping; an inverted filter of graded, crushed rock may have to be provided at the base to control piping where the bed sediments are fine grained.
- The core, transition, and outer supporting rockfill zones on the basin side would be raised to mean sea level simultaneously in lifts of up to, perhaps, 2.5 m, ensuring that the rockfill elevation does not lag the other zones by more than twice that lift.
- Raising of all zones in small lifts would continue, maintaining adequate protection of the core and transition zones from wave attack.
- The protective layers of rip-rap or armor stone would be placed to suit, either simultaneously or upon completion of the other zones.

Since the materials in the dike would be end dumped into moving water, the actual embankment slopes achieved will depend to some extent on the placing conditions and the current velocities. Quantity estimates could be based on a slope of 1.75 : 1 for current velocities up to 3 m/s. The slopes to be adopted will depend on the results of model tests carried out to determine the performance of dumped fill under the prevailing tidal conditions. The thickness of the core zone for an ebb generation development would depend upon the net operating head; for a double-effect development the full tidal range should be used.

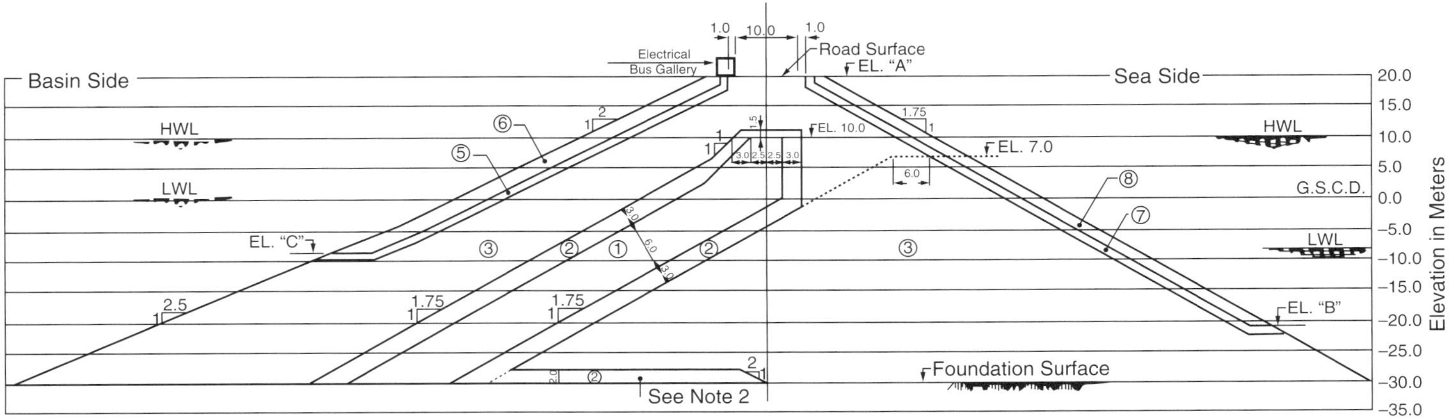

Notes

1. Proposed access dike section assumes a bedrock foundation or competent overburden. Soft or loose sediments will be dredged or effectively displaced by advancing rockfill.
2. In locations where access dike foundation is composed of fine-grained sediments, an inverted filter will be required under the rockfill to control piping.
3. HWL refers to higher high-water large tide plus 1.22 m (50-year return period). LWL refers to lower low-water large tide minus 0.76 m (50-year return period).
4. Elevations, water levels and sizes of armor bedding and armor are site specific.

Legend

① **Core zone** (sand and gravel)
② **Transition zone** (crushed rock or equivalent)
③ **Rockfill zone D_{50} plus 200 mm** (quarry run)
④ **Omitted in access dike**
⑤ **Armor bedding zone, basin side** (see table for sizes)
⑥ **Armor zone, basin side** (see table for sizes)
⑦ **Armor bedding zone, sea side** (see table for sizes)
⑧ **Armor zone, sea side (dolos)** (see table for sizes)

Construction Sequence

1. Prepare foundation of access dike by removing soft and loose sediments.
2. Construct Zones 3 to elevation 7.0 m. This segment need not be built up in lifts, except where such procedure is necessary to provide the required side slopes. Elevation 7.0 m does not provide access at all times.
3. Construct Zones 1 and 2 simultaneously in lifts not exceeding 2.5 m in thickness, up to a maximum elevation of zero G.S.C.D., starting from the completed rockfill embankment and working toward the basin. The basin side rockfill Zone 3 should advance behind Zones 1 and 2 with a maximum height differential of 5 m.
4. Above elevation zero G.S.C.D., construct Zones 1 and 2 simultaneously in lifts not exceeding 1 m in thickness and maintaining a maximum differential of 1 m. Zone 2 on the sea side should be maintained at the same elevation as, or higher than, Zone 1. The sea side Zone 3 should be constructed in 2 m lifts and maintained at the same level as, or higher than, Zone 2.
5. Zone 5, 6, 7, and 8 may be constructed either simultaneously with or upon completion of the other zones.
6. Zone dimensions indicated on the drawing are minimum. Increases in lift thickness greater than indicated may be considered provided the minimum zonal dimensions are obtained and no significant segregation occurs in placement of the material.

Figure 8.12 Typical cross section of proposed access dike for single-effect Bay of Fundy scheme.

8.5.2 Closure Dike

As the construction of the barrage proceeds across an estuary to form a tidal basin, the area of flow is gradually decreased and a corresponding reduction in total discharge into and out of the tidal basin occurs. The basin water level is no longer able to change as fast as the seaside tide. The resulting water level differential causes dramatic increases in flow velocity and also causes the basin tide to lag behind the open-sea tide and, ultimately, to reduce in range when the basin can no longer be completely filled or emptied before the sea tide turns. Completion of the barrage is usually referred to as *closure*. Final closure of the gap in the tidal barrage, incurring the hydraulic problems associated with increasing heads and tidal velocities, is one of the critical aspects of tidal plant construction. Particular attention must be given to the selection of construction procedures so as to minimize the potentially hazardous and costly aspects of this activity. At all stages of closure, the elements of the barrage must be able to withstand the hydraulic loads arising from these head differences and velocities.

Obviously, the closure problem is severe at sites where the tidal range is large enough to make tidal power attractive. The caisson method of construction is particularly useful in the solution of the closure problem since the large number of "holes" in the barrages, that is, the turbine and sluiceway passages, can be used during construction as diversion waterways. Very careful scheduling of the various stages of constructon is necessary to keep the development technically feasible and to keep costs to a minimum.

A mathematical model, to predict water velocities as a tidal basin is closed, is a very effective tool to guide preliminary dike designs and construction sequences. The important parameters in determining closure velocities are: tidal range, closure width, barrage height, basin storage area, as well as the effective number of sluiceways. Because of the importance of this operation, hydraulic model studies would be required before final design criteria could be selected. For preliminary estimates, however, the computer model with a number of simplifying assumptions can be used. The computed tidal current variations during construction of the closure dike at a proposed site in the Bay of Fundy are illustrated by Fig. 8.13.

The dike design is almost fully determined by the placement conditions. Velocities during final closure may be in the order of 10 m/s so that rock fragments up to $D_{50} = 1.6$ m or a weight of about 6 tonnes would be required to withstand the force of the maximum currents under spring tide conditions. Larger sizes would be needed for higher velocities. Computed average maximum velocities during closure for a proposed barrage construction sequence in Cumberland Basin is shown in Fig. 8.14.

A typical cross section of a closure dike is given in Fig. 8.15. Considering tidal plant requirements, the closure dike would not have to be raised high enough to provide continuous access at all times across the barrage. For purposes requiring such access, such as regional communications, the design could be modified.

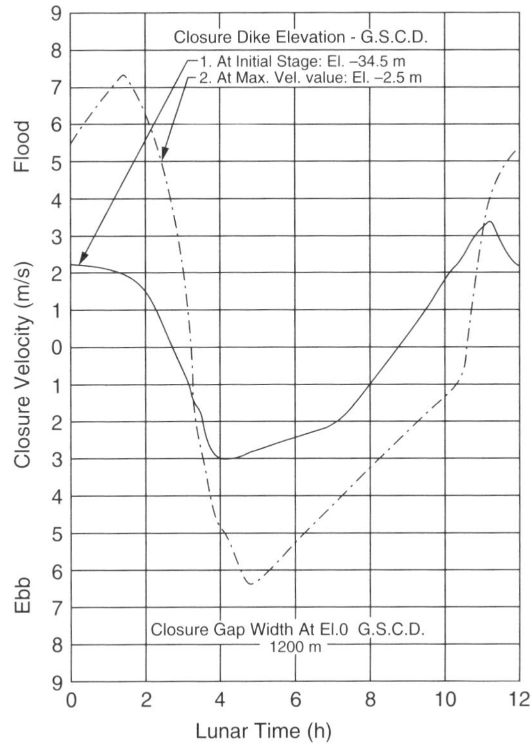

Figure 8.13 Tidal current variations during construction of closure dike.

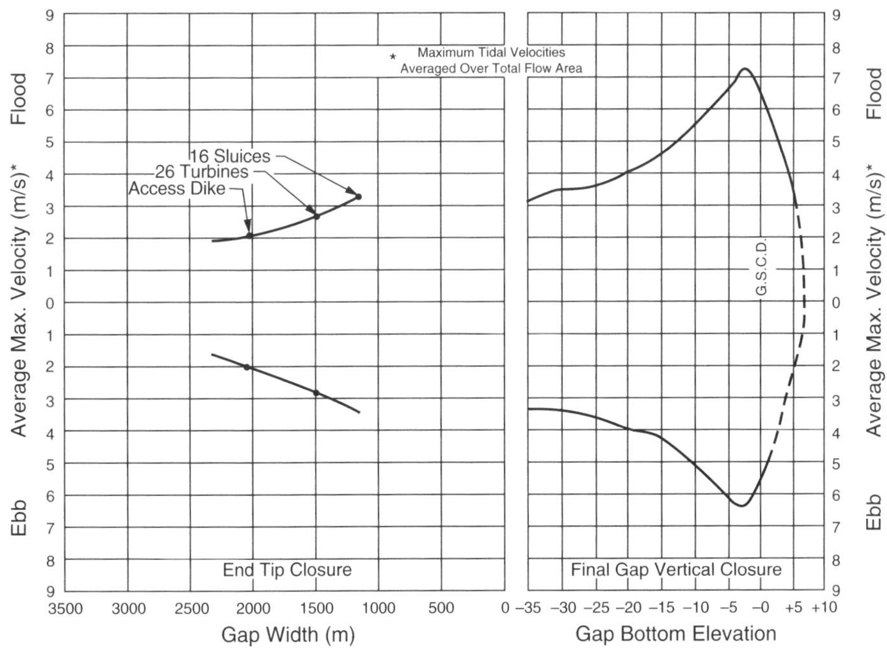

Figure 8.14 Computed average maximum closure velocities.

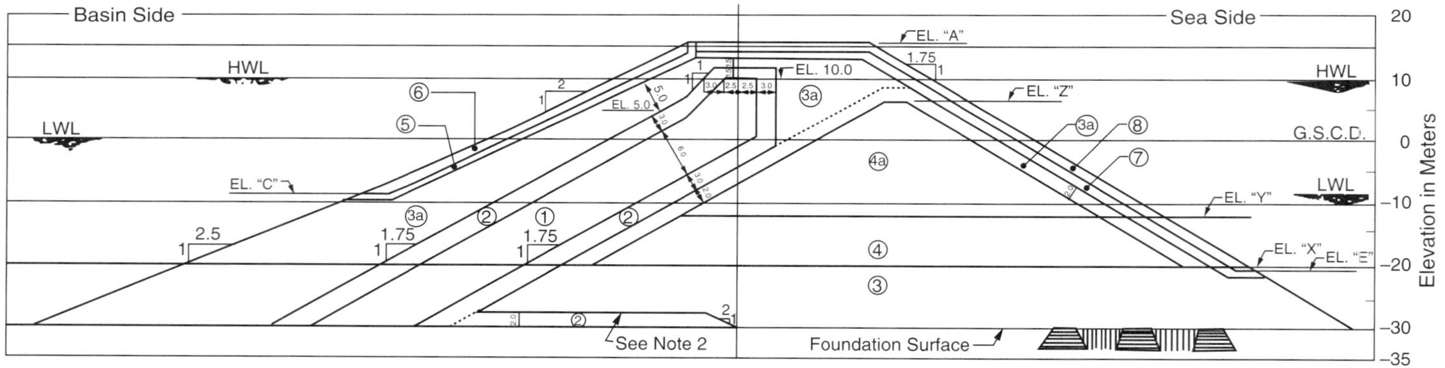

Notes

1. Proposed closure dike section assumes a bedrock foundation or competent overburden. Soft or Loose sediments will be dredged or effectively displaced by advancing rockfill.
2. In locations where the closure dike foundation is composed of fine-grained sediments, an inverted filter will be required under the rockfill to control piping.
3. The design of the rockfill embankment is controlled by water velocities developing upon closure. The velocity profiles used in this design are those computed for each specific site. Placement of the rockfill embankment will be by the vertical closure method.
4. HWL refers to higher high-water large tide plus 1.22 m (50-year return period). LWL refers to lower low-water large tide minus 0.76 m (50-year return period).
5. Elevations, water levels and sizes of armor bedding and armoring (dolos) are site specific.

Legend

① **Core zone** (sand and gravel)
② **Transition zone** (crushed rock or equivalent)
③ **Rockfill zone D_{50} = 200 mm** (quarry run)
③ ④ ④a **Rockfill zones** (see Table for sizes)
⑤ **Armor bedding zone, basin side** (see table for sizes)
⑥ **Armor zone, basin side** (see table for sizes)
⑦ **Armor bedding zone, sea side** (see table for sizes)
⑧ **Armor zone, sea side (dolos)** (see table for sizes)

Construction Sequence

1. Prepare foundation of closure dike by removing soft and loose sediments.
2. Construct Zones 3 and 4 to El. X and Y respectively. These segments need not be built up in lifts, except where such procedure is necessary to provide the required side slopes.
3. Proceed with Zones 4a from El. Y to El. Z using a lift thickness of 5 m
4. If access is required, Zone 3a rockfill adjacent to Zones 4 and 4a rockfill may be placed to elevation +8 m. Construction in lifts is not required. El. +8 m does not provide access at all times.
5. Construct Zones 1 and 2 simultaneously, in lifts not exceeding 2.5 m in thickness, up to a maximum elevation of zero G.S.C.D. starting from the completed rockfill embankment and working toward the basin. The basin-side rockfill Zone 3a should advance behind Zones 1 and 2 with a maximum height differential of 5 m
6. Above elevation zero G.S.C.D. construct Zones 1 and 2 simultaneously in lifts not exceeding 1 m in thickness and maintaining a maximum differential of 1 m Zone 2 on the sea side should be maintained at the same elevation as, or higher than, Zone 1. The sea side Zone 3a should be constructed in 2 m lifts and maintained at the same level or higher than Zone 2.
7. Zone 5, 6, 7, and 8 may be constructed either simultaneously with or upon completion of the other zones.
8. Zonal dimensions indicated on the drawing are minimum. Increases in lift thickness greater than indicated herein can be considered provided the minimum zonal dimensions are obtained and no significant segregation occurs in placement of material.

Figure 8.15 Typical cross section of proposed closure dike for single-effect Bay of Fundy scheme.

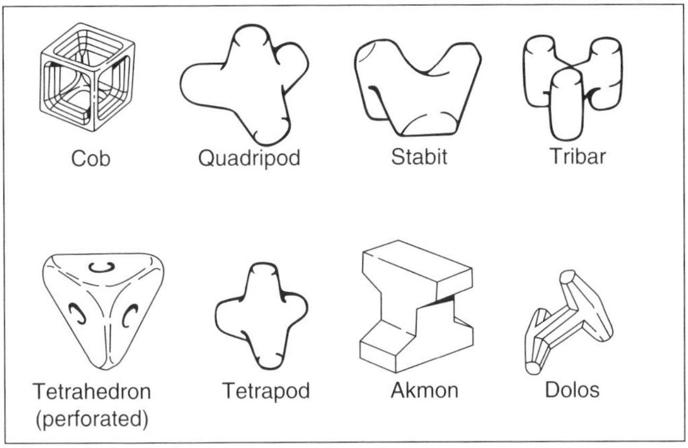

Figure 8.16 Shapes of precast concrete armor units.

The type of protective layer would be based on the anticipated wave runup, wind setup, and tidal range. The main armoring could be by means of rip-rap (graded rock with a D_{50} size chosen to suit the design, wave climate, and the slope of embankment face) if there is an adequate local source available, or by precast concrete units of special design, some of which are shown in Fig. 8.16.

8.5.3 Availability of Construction Materials

Locating sources of indigenous construction materials, such as impervious materials, granular materials and concrete aggregates, rockfill materials, and armor stone, requires field investigations for which aerial photo interpretation could provide much useful information. If quantities of materials must be hauled to the site over long distances, the cost of development could rise substantially since huge quantities are required.

8.6 CONSTRUCTION SCHEDULES

Construction scheduling is a very important aspect in keeping the costs to a minimum. Apart from the mobilization period, final site investigation and final design could require a minimum of 3 years. If a construction sequence can be developed that would permit operation of a number of machines even before final enclosure is accomplished, the revenue from the output could materially offset interest during construction, which makes up a substantial amount of the final cost to the time of commissioning of the entire plant.

Two tasks fundamental to a full and final study are the development and operation of a model of the estuary, possibly a hybrid model, and numerical

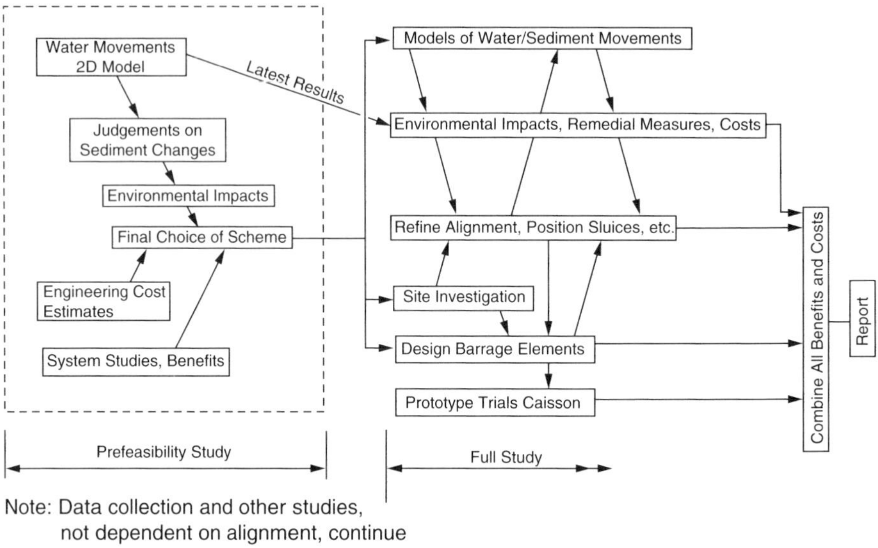

Figure 8.17 Decision flowchart for studies dependent on barrage alignment.

models of sediment movements. Such modeling would also be central to much of the later environmental work.

The final study and design would be an iterative process through which the barrage design is gradually refined. For example, Fig. 8.17 presents a decision flowchart for studies dependent on barrage alignment as developed by the Severn Barrage Committee (1981a).

CHAPTER 9

ELECTROMECHANICAL EQUIPMENT FOR A TIDAL PLANT

9.1 INTRODUCTION

The turbine generating equipment selected for a site is probably the greatest single factor affecting the direct costs of the development. Electromechanical equipment will generally account for between 45 and 55 percent of these costs and indirectly affect a further 30 to 35 percent since the powerhouse is largely dependent on the main dimensions of the machines, and, in addition, the number and size of the sluices are dependent on the sluicing capacity of the turbines. On the other hand, the output of tidal power plants depends entirely on the turbine concept chosen and on the discharge and efficiency of the turbines.

In the three decades following the major effort initiated to develop equipment specifically suited to the tidal power development contemplated for the Rance Estuary (France), emphasis on technological development of low-head, axial-flow turbines has been directed to conventional river hydro. However, some attention has been given specifically to turbine requirements for tidal power developments with their constantly varying heads and the large flows under which they must operate, particularly in view of the high proportion of total development costs contributed by the turbogenerating equipment. Nevertheless, basic information at this time must still be derived from conventional river hydroelectric plants.

Low-head power equipment has already been shown to be adaptable to tidal-electric plants where potential heads of 5 to 13 m may be expected. Such

Elements of Tidal-Electric Engineering. By Robert H. Clark

a range is covered by the axial-flow design with its relatively high specific speed. Because of their hydraulic and cavitation performance and the economy of large diameter, the preference for relatively low rotational speed turbine units are well established.

9.2 SPECIFIC REQUIREMENTS FOR TIDAL GENERATING EQUIPMENT

Generating equipment for tidal-electric plants must take into account the following factors:

1. The available head is low and continuously changing requiring turbines with features to provide satisfactory operating characteristics and efficiency under such conditions.
2. The low head requires high volume water discharge resulting in large dimensions for water passages and high unit kilowatt equipment cost.
3. The cyclical operation (the starts and stops in synchronism with the tides) imposes greater stresses on the generating equipment than would occur in a conventional river hydroelectric plant.
4. The number of units in a major development will be very large [100 and more are being proposed for sites in the Bay of Fundy (Canada) and the Severn Estuary (Great Britain)].
5. Unit dimensions must be kept as small as possible in view of the consequent overall size and civil works cost of the powerhouse as resulting from the great number of units generally involved (comparative dimensions are shown in Fig. 9.1);
6. The materials are exposed to the corrosive action of seawater and therefore must be carefully selected and protected.
7. Overall efficiency of generation may be assigned a lower priority in view of the ample supply of water involved.

Since the economy of a tidal power scheme is significantly influenced by the number of turbogenerators, savings may be realized by using units with capacities as large as possible. However, increasing the size of the turbogenerator reduces its rotational speed, which, for tidal power units, is kept low in any case in order to increase its operational head range. Moreover, tidal energy exploitation is associated with relatively low heads, which also means low rotational speeds. The relationships of power, head, and speed are expressed as follows:

$$P \sim D^2 \quad (9.1)$$

$$n \sim \frac{1}{D} \quad (9.2)$$

$$n \sim \sqrt{H} \quad (9.3)$$

where P = output
D = turbine runner diameter
n = turbine speed
H = rated head

As a consequence, the generator will have definite limits as far as size and design output are concerned (Seoni, 1977) since the generator output is defined by the following relationship:

$$P \sim K \times D_r^2 \times Ln \quad (9.4)$$

where P = output
K = compactness factor
D_r = rotor outside diameter
L = rotor length
n = generator speed

9.3 TYPES OF TURBINES

As noted previously, the choice of concept for the hydraulic machinery is of the greatest importance for the economy of a tidal power plant, which must exploit energy under low to very low heads. The only turbine types adaptable to such plants are the axial-flow, high specific-speed turbines, of which there are four basic types: Kaplan, tube, bulb, and straight flow. The governing dimensions in terms of runner diameter of these four types are shown in Fig. 9.1.

Simplicity and compactness are prime requirements to make optimum use of the caisson concept for barrage construction, which has proven so successful in building dikes, oil-drill platforms, and similar structures (Derrington, 1979). However, these requirements are not fulfilled by the vertical Kaplan turbine, nor by the tubular turbine (Fig. 9.1). In the former, which is a vertical-axis machine, losses are incurred in directing the water through bends in the intake, distributor, and draft tube or water passage, downstream of the runner. Moreover, the vertical-axis arrangement requires relatively deep water and so may increase construction costs. The tubular turbine, although overcoming generally the aforementioned problems associated with the vertical Kaplan turbine, introduces problems related to critical speed because of the long drive shaft that operates in cross flows and thus would be subjected to complex bending stresses as well as torsion. For small-scale tidal power schemes, however, the tubular turbine, that is, a geared bulb-turbine assembly with a horizontal or inclined shaft (Fig. 9.2), could be economically attractive.

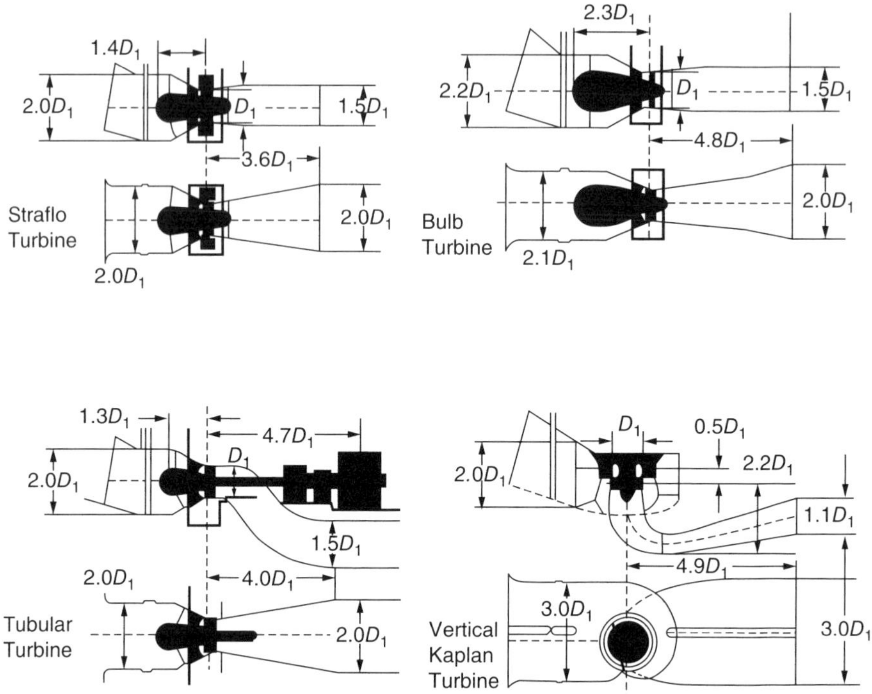

Figure 9.1 Overall dimensions for low-head turbine designs.

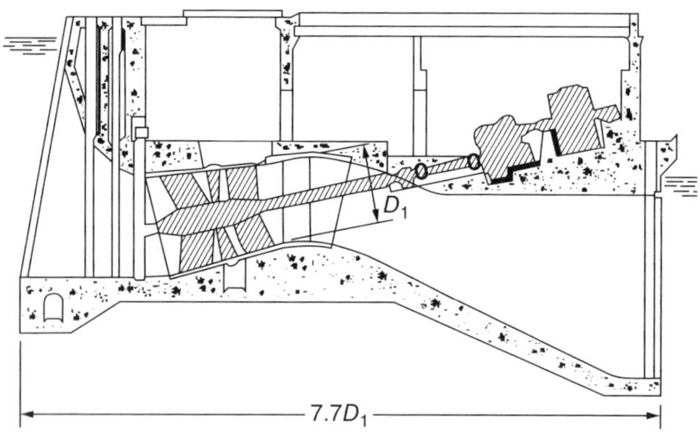

Figure 9.2 Cross section through a typical powerhouse for a slant-axis, tubular turbine.

9.3.1 Bulb Turbines

The bulb turbine (Figs. 9.1 and 9.3) operates in a straight water passage with the generator enclosed in a bulb. The assembly is made up of stay vanes, adjustable wicket gates, and variable-pitch runner vanes on a horizontal shaft,

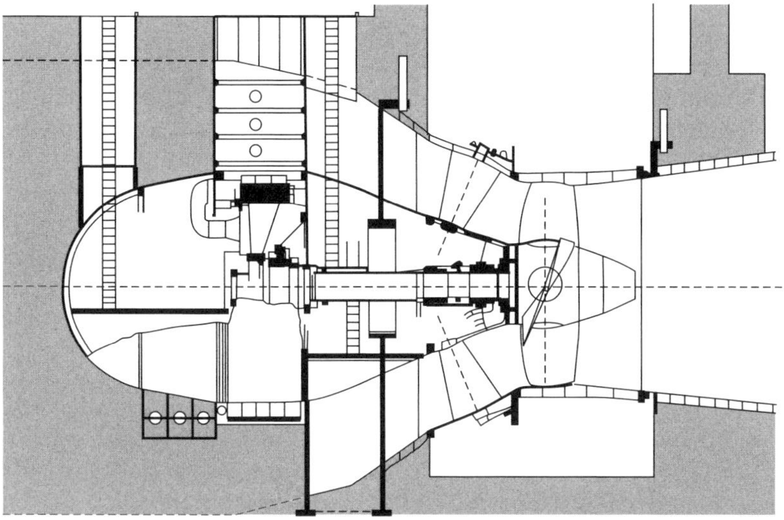

Figure 9.3 Cross section of bulb turbine with direct-driven generator for the Racine hydropower plant (United States). (From Miller, 1978.)

thereby providing *double regulation*, that is, regulation of the flow by both the wicket gates and runner blades, and improving the overall efficiency. The stay vanes provide the structural support from which the turbine unit is cantilevered. The generator is enclosed in a steel bulb within the water passage. Since the draft tube is essentially straight and horizontal, a greater percentage of the kinetic energy exiting from the runner is retrieved than is achieved by the draft tubes of the vertical Kaplan and tubular turbines. The operation and efficiency of the draft tube plays a significant role in the overall efficiency and performance of the turbine. The straight and essentially horizontal waterway achieves attractive savings in the volume and cost of the civil works.

This design, of which the outstanding installation in tidal power application is at the Rance development in France (refer to Chapter 15), made its first appearance in the mid-1930s in a small run-of-river plant. Major developments took place in the postwar years in France, promoted to a large extent by a desire to create a suitable unit for application in tidal power development. The bulb turbogenerating unit is characterized by improved hydraulic efficiency, greater discharge capability, and greater power output for the same runner size than other designs. Moreover, it reaches higher specific speeds than the Kaplan turbine and so can be used to exploit even lower heads.

The bulb turbine has been a significant advance in the field of low-head hydro developments, enabling the exploitation of heads between about 1.5 to 20 m. Its characteristics are well developed and almost all recent installations for large, low-head, river schemes have used bulb turbines. Table 9.1 lists some of the major low-head installations since 1967. The normal arrange-

TABLE 9.1 Some Major Low-Head Installations Since 1967

Country	Location	Turbine Diameter (m)	Speed (rpm)	Generator Size (MW)[a]	Rated Head (m)	No. of Units	Date of Commissioning	Type of Turbine	Method of Cooling Generator
Austria	Altenworth	6.0	103.4	38.9	14	9	1976	Bulb, double regulated	Pressurized air
Canada	Jenpeg, Manitoba	7.5	62.1	31.1 @ 60 Hz	10.7	6	1977	Bulb, double regulated	Direct water
	Annapolis Royal	7.6	50	20 @ 60 Hz	5.5	1	1984	Straflo, single regulated	Air
France	Peage-de-Roussilon	6.25	93.8	40	12	4	1977	Bulb, double regulated	Pressurized air
	Caderousse	6.25	93.75	31.5	9.5	2	1975	Bulb, double regulated	Pressurized air
	Caderousse	6.9	93.75	31.5	9.5	4	1975	Bulb, double regulated	Pressurized air
	Sauveterre	6.9	93.8	33	9	2	1973	Bulb, double regulated	Pressurized air
	Avignon	6.25	93.8	30	10	4	1973	Bulb, double regulated	Pressurized air
	La Rance	5.35	93.75	10	5.75	24	1966/7	Bulb, pump-turbine	Pressurized air
Germany	Iffezheim	5.8	100	28.3	11.7	4	1977	Bulb, double regulated	Pressurized air
United States	Ozark	8	60	25 @ 60 Hz	10.6	5	1970	Tubular, double regulated	Air
	Harry S. Truman	6.45	94.7	31.5 @ 60 Hz	24	6	1976	Tubular, pump turbine	Air
	Rock Island	7.4	85.7	54 @ 60 Hz	12.1	8	1978	Bulb, double regulated	Pressurized air
	Racine	7.7	62.1	24.6 @ 60 Hz	6.2	2	1981	Bulb, double regulated	Pressurized air
	Vidalia	8.2	52.2	24 @ 60 Hz	4.9	8	1989	Bulb, single regulated	Pressurized air

[a] At 50 Hz unless shown otherwise.

ment is for the axis of the machine to be horizontal with the bulb housing the generator located upstream of the turbine runner, which is directly coupled to the generator.

The compactness of the bulb turbine, however, brings with it some drawbacks. For large machines such as the turbines for the Racine hydroelectric plant in the United States (Fig. 9.3), where the operating head is relatively small, the rotational speed of the turbines must be kept low to avoid cavitation damage. This requirement poses a problem since the generator diameter increases as the chosen speed is reduced. Because the generator diameter defines the diameter of the bulb, increases in the generator size results in an increase in the length of the convergent section of the waterway in an effort to avoid higher energy losses that would result from the sharper change in the direction of flow past the bulb. Moreover, this disproportionate lengthening of the generator increases its cooling difficulties.

In addition, with the generator housed in the upstream bulb, which has to be as small as possible for hydraulic and economic reasons, the electrical output is concentrated in a small diameter so that the natural inertia of the unit is low. The inertia or flywheel effect (WR^2) of a turbogenerator set may be attributed primarily to the mass of the generator rotor (Mosonyi, 1987). The turbine intake must be very carefully designed to avoid unacceptable fluctuations in generator output. Tidal power units are particularly subject to rough operation due not only to pressure fluctuations caused by ocean waves but also because a tidal, turbogenerating unit seldom operates in the optimum hydraulic range.

Nevertheless, to this point in time, the bulb turbine has had the advantage of experience and expertise in its design and use, and it is generally chosen as the machine on which to base tidal output optimization and civil works designs (Bosc, 1997). Moreover, it is adaptable to operation in six possible modes: as a turbine, a pump, and an orifice in both directions, the latter at up to 50 percent of the discharge capacity.

9.3.2 Straight-Flow (Straflo) Turbine

The concept of mounting the generator rotor on a rim fixed peripherally to the runner blades (also known as a rim-generator turbine, Figs. 9.1 and 9.4) was patented in 1919 by Harza. Subsequently, such machines were developed and built by Escher-Wyss during the period 1936–1953, but limited to relatively small runner diameters of the order of 1 m. That firm resumed development of the seals for large-diameter machines and, in 1984, the Annapolis Tidal Generating Station in Nova Scotia, Canada, was commissioned. This station contains one 7.6-m-diameter Straflo turbine with a maximum rated output of 19,600 kW under a rated head of 5.5 m.

The Straflo turbine does not have the obstruction of a large bulb containing the generator in the waterway so that hydraulic losses are lower. The absence of the "bulb" results in shorter, less expensive, and more hydraulically

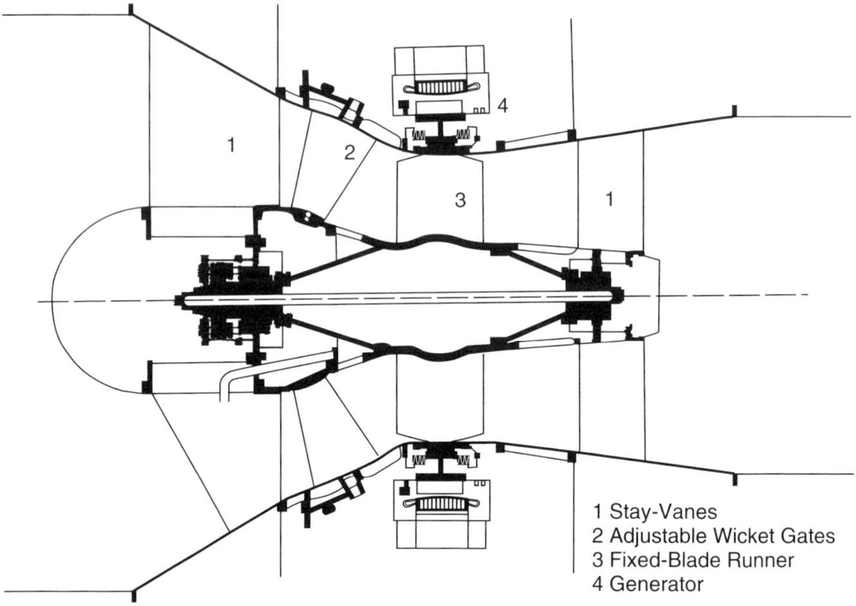

Figure 9.4 Typical cross section of a Straflo turbine. (From Miller, 1978.)

efficient water passages. The concept of using the turbine as the generator rotor relaxes the physical constraints affecting the number of poles and the space available for cooling. In further contrast to the bulb turbogenerator unit, there is no restriction on the diameter of the Straflo generator so that the inertia of the latter is nearer that of the conventional vertical shaft Kaplan unit, and the regulation difficulties inherent with the bulb unit, with its abnormally low inertia, are avoided. Moreover, since the Straflo generator is outside the turbine water passages, that is, it is not in a steel shell in the water, it is more easily accessible for maintenance. Both turbine and generator are in the same vertical plane and so can be handled by the same crane, substantially reducing the area to be covered by cranes and, consequently, the size and cost of the powerhouse.

A comparison of bulb and Straflo turbine arrangements under the same rated head is shown by Fig. 9.5. This comparison does not indicate the possible increase in power output of the Straflo unit.

The large-diameter rotating rim initially presented difficult sealing problems that have now been resolved by an ingenious solution. The seals are required to exclude both water and suspended sediments from the generator under the large difference in water pressure between the top and bottom of the rim, as well as radial movements caused by temperature changes and by the deflections caused by changes in loads in the runner blades.

The largest Straflo turbogenerator built to date is the 7.6-m diameter, 20-MW machine commissioned in 1984 in the powerhouse constructed in an

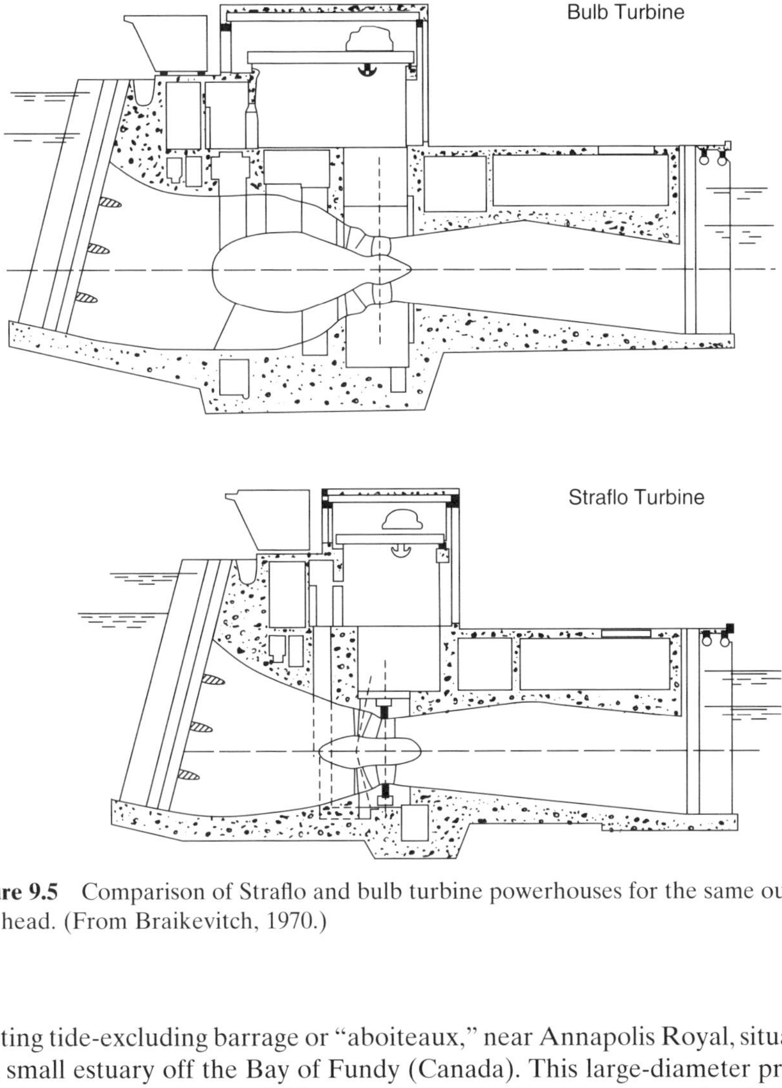

Figure 9.5 Comparison of Straflo and bulb turbine powerhouses for the same output and head. (From Braikevitch, 1970.)

existing tide-excluding barrage or "aboiteaux," near Annapolis Royal, situated in a small estuary off the Bay of Fundy (Canada). This large-diameter prototype, known as the Annapolis Demonstration Project, operates under the tidal head fluctuations in the estuary, and its performance is of particular relevance to tidal schemes (refer to Chapter 15). During its ensuing period of operation from 1984 through 1996, the relatively few problems encountered have been associated with auxiliary equipment rather than the turbine design as such. The turbine runs very smoothly and quietly and has operated with a high level of availability since it entered commercial service. It provides rated output at rated head and has required relatively little maintenance (Clark, 1997).

With the evolution of designs to improve thrust and guide bearing supports to enhance further the economics of this type of turbine, the hydraulic performance of the rim generator turbine is now very similar to that of an equivalent bulb turbine.

9.3.3 Methods of Regulation

There are three methods of regulating the output of bulb turbines. These are: (1) varying the angle of the runner blades, (2) varying the opening of the guide vanes forming the distributor, and (3) with the generator connected to the national grid via a direct-current (dc) link, varying the speed and torque by altering the strength of the generator rotor field. These can be used singly or in combination. However, for the Straflo turbine, at least at this stage in its development, the runner blades must remain fixed.

In those instances where the variations of head or flow through a turbine are low, the need for regulation is also low, such as for run-of-river schemes. This is not the case for tidal schemes where the head varies continuously and regulation becomes necessary to enable the turbine to maintain good efficiency over the range of operating conditions. However, installing adjustable runner blades and/or distributor vanes carries a cost penalty that may be justified by the increased energy output resulting from the increased efficiency. For turbines capable of generating in both directions of flow, both forms of regulation, known as *double regulation*, would be used to achieve reasonably effective output in the two directions.

If only one method of mechanical arrangement were appropriate, adjustable runner blades offer a slight cost advantage since only 4 parts of the turbine move relative to each other compared with, perhaps, the 20 vanes making up the distributor, so that sealing and associated maintenance problems are reduced. However, with only the runner blades variable (distributor vanes fixed), a separate downstream gate is necessary for a tidal-electric unit to control starting and stopping of the water flow past the turbine for both the generating cycle and reverse sluicing cycle (assuming ebb-flow generation). The gate would also have to be shut quickly at any time should a mechanical or electrical fault develop and, therefore, designed to close against the maximum runaway flow, that is, the flow causing runaway speed. Although such an arrangement would present its own maintenance problems, these would probably be less severe than those associated with movable distributor vanes. However, it should be noted that, because of its high inertia, the runaway speed for a Straflo turbine generator should not be particularly dangerous, particularly under the magnitude of rated heads for tidal plants.

Figure 9.6 illustrates the relative variation in annual energy output from a potential tidal development under different regulation modes for a 60-MW synchronous generator capacity limit. The dashed line indicates the output from a variable-speed machine with variable guide vanes.

In addition to its regulating function, a distributor ring with movable guide vanes would normally control both starting and stopping of the turbine at each end of the generation part of the tidal cycle and also reverse sluicing through

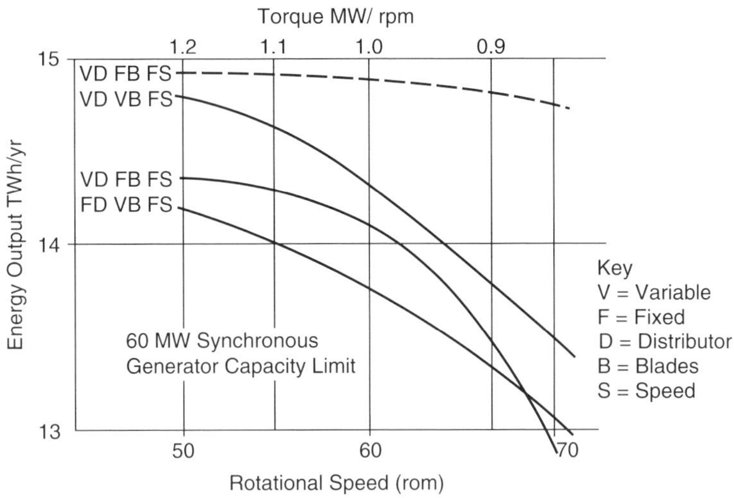

Figure 9.6 Energy output for fixed and variable speed generation. (Adapted from Severn Barrage Committee, 1981a.)

the turbine during the refilling of the basin in ebb generation schemes. Such a variable distributor could also shut down the turbine in an emergency as a result of a mechanical fault such as an overheated bearing or an electrical fault causing the load on the generator to be shed. Since the distributor and operating mechanism is independent of the rest of the turbine, it should provide a reliable method of emergency shutdown.

Effective extraction of energy by a tidal-electric plant places more emphasis on high unit power (i.e., power developed from a given runner design of unit diameter under unit net head) than on extreme values of efficiency. Flows used to produce tidal energy are more effectively employed at the higher heads occurring through the cycle and, because of the greater capacity of a tidal-electric plant to use these flows, can increase production. In conventional hydroelectric plants, variations from rated conditions (generally near average head) occur more rarely so that generator output is set for rated conditions and accepted as a limitation during high head conditions.

The available head on a tidal power plant for ebb generation and those features affecting turbine design are illustrated by Fig. 10.1. The cycle of performance of a tidal power unit is perhaps best visualized by observing the trace that it would follow on a "hill" chart. Traces for maximum efficiency and maximum power are shown by the heavy lines in Fig. 9.7. Figure 9.7*a* shows the characteristics for a fixed runner blade and variable distributor configuration, and Fig. 9.7*b* illustrates the performance for a turbine when both distributor and turbine blades are variable in angular setting.

Rated head, that is, unity ratio on the hill curves, is the head at which the generator is capable of delivering its limiting capacity. In utilizing such unit-

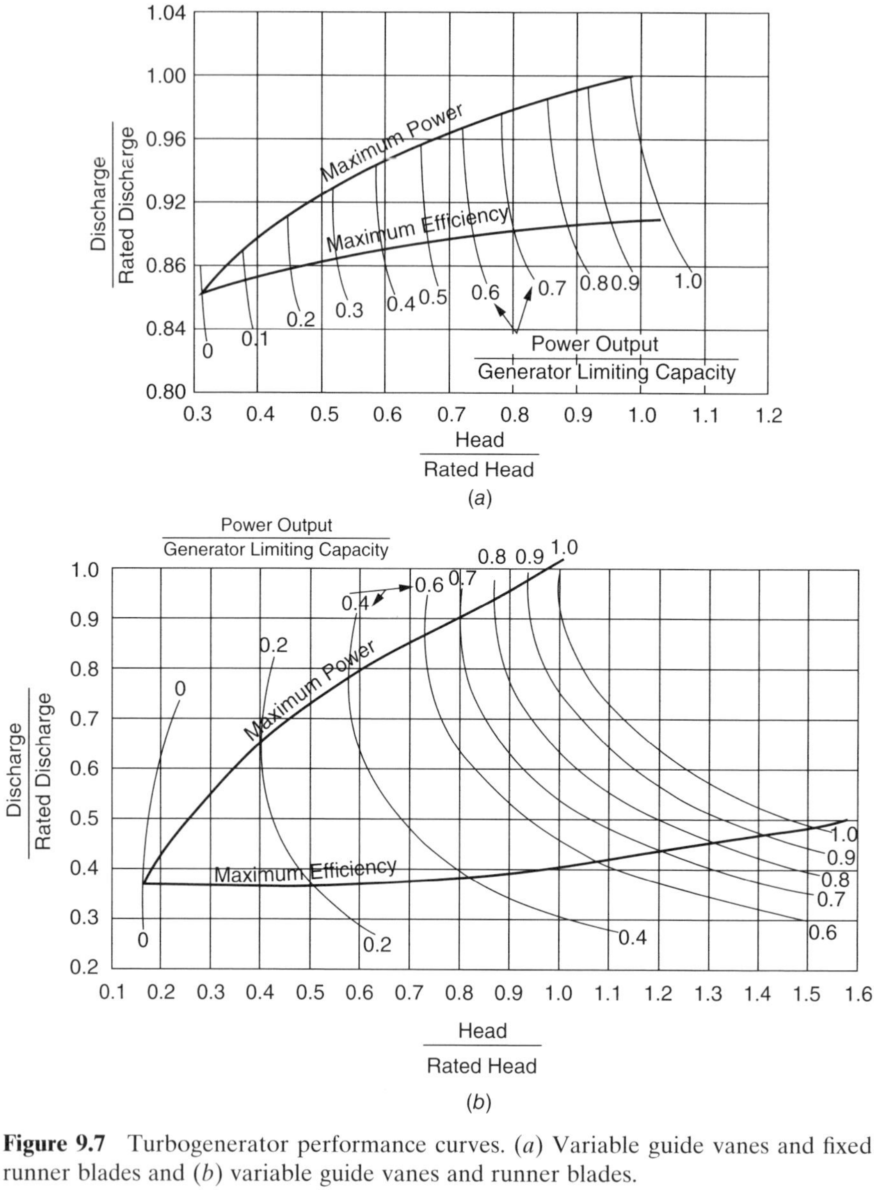

Figure 9.7 Turbogenerator performance curves. (*a*) Variable guide vanes and fixed runner blades and (*b*) variable guide vanes and runner blades.

ized characteristic curves developed from manufacturer's data, care should be taken to ensure that the efficiencies also reflect the intake and draft tube losses. For fixed-blade installations, it is assumed that the units will be operated along the maximum power and limiting power curves at all stages of the tide, resulting in a deterministic relationship between head and discharge. This assumption introduces a slight error in the assessment of maximum production, amounting to a slight loss of about one-half of one percent of the annual energy production in comparison with the true optimum; the error introduced by this assumption can, therefore, be approximately taken into account in computing annual revenues. Operation of variable pitch blade turbines for maximum energy production would be simulated in accordance with the more complex pathway through the turbine “hill chart” determined by introducing the calculus of variations to define an optimal relationship for maximum energy or benefit between discharge, basin level, and time during the generating period.

The angle of the blade setting must be selected for a fixed runner blade turbine. This, of course, affects the relative power output. Increasing the angle may permit either the selection of a significantly smaller diameter machine or higher discharges and power output if the diamater were held constant. Since unit efficiency is not a critical factor, a fixed-blade angle of 27.5°, that is, the *setting angle* or the angle of the runner blade with the plane of rotation (Fig. 9.8) and a resulting peak efficiency of 83 percent were assumed for the Bay of Fundy reassessment studies. A fixed-blade configuration has much to commend it for cost and simplicity; the added zone of operational convenience provided by adjustable blades is fairly limited for a tidal-electric plant. On the other hand, it should be kept in mind that, for a blade angle that is less than optimal, the water leaving the blade would have to make a sharp change in direction, which could create shearing forces sufficient to damage or kill tender fish, such as shad, although such mortality would be lowest when the turbine runs near peak efficiency. Also, operation at limiting distributor guide vane angles with fixed runner blades may bring in operational stability limitations.

Double-regulated turbines, that is, with both turbine blades and distributor vanes variable, are normally chosen where the range of operating heads and flows is wide. At any head, both the distributor guide vanes and the runner blades can be adjusted to give the maximum efficiency for the required flow of water. Thus, for tidal schemes, fixed-speed turbines give the maximum output when double regulated (refer to Fig. 9.7). The extra cost of double regulation

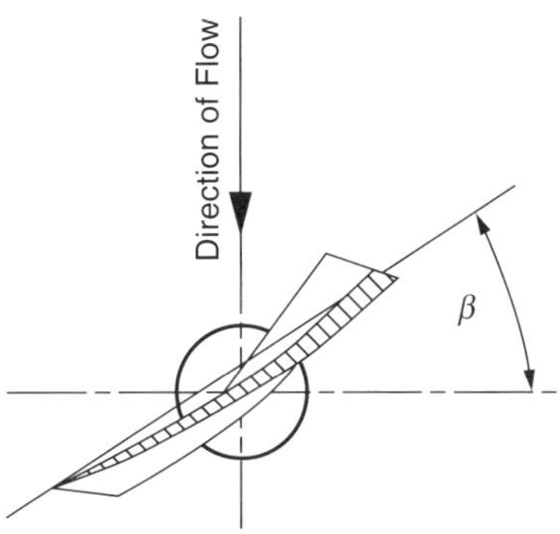

Figure 9.8 Runner blade setting angle.

would have to be weighed against the value of extra energy produced. In the investigation of the Bay of Fundy sites, slightly lower costs of energy were estimated for variable-pitch than for fixed-blade runners. For example, at the Cobequid Bay site, that is, B-9 on Fig. 16.16, the increase in energy produced by variable-pitch blades would be in the range of 3 to 4 percent—an increase compatible with the extra cost of selecting variable-pitch units for use in the powerhouse since the mechanical components represent some 25 percent of the total capital cost of a tidal power development, and it is estimated that the variable-pitch unit would cost about 12 to 14 percent more than a fixed-blade turbine.

For schemes involving two-way generation or pumping, or both, double-regulated turbogenerators are essential if reasonable efficiency is to be achieved when pumping or when generating in the reverse direction. In each case, the optimization of turbine design and of the method of operation are very complex and have to take into account external factors, such as variations in the value of energy with time of day.

Recent developments in thyristor technology mean that it is feasible to permit the turbine to run at variable speeds. The turbine would not be governed so as to maintain synchronous speed, but the speed would be allowed to vary with the flow through the turbine and the head across it, and with the load on the generator. In other words, the output of normal alternating-current (ac) generators would be rectified, the power transmitted as dc to a convenient location and then inverted back to ac at the precise frequency of the grid. For tidal schemes, this would have advantages in increasing the overall energy output and in eliminating difficulties in synchronizing or maintaining synchronous speed as a result of wave attack during severe storms. Figure 9.6 compares the output with other regulation configurations. Although the rectifier and inverter stations would impose losses of, perhaps, 1 to 1.5 percent of the energy generated, line losses for dc transmission would be lower. The relative costs of ac and dc transmission would depend upon the proposed line capacity and length.

9.3.4 Specific Speed

A considerable number of large-diameter bulb turbines have now been installed (refer to Table 9.1), and there is available a great deal of information relating to characteristics and performance. However, as noted previously, there is limited experience of the effects of cyclic loading and stressing of bulb units under tidal conditions and no such experience with large units (>8 m diameter).

Figures 9.9 and 9.10 indicate the relationships of computed specific speed with head and with unit power output for bulb turbines. These three turbine characteristics are useful in the selection of the appropriate size and output for a given head. The specific speed of a turbine is that speed [revolutions per minute (rpm)] at which a turbine of homologous design would operate if the

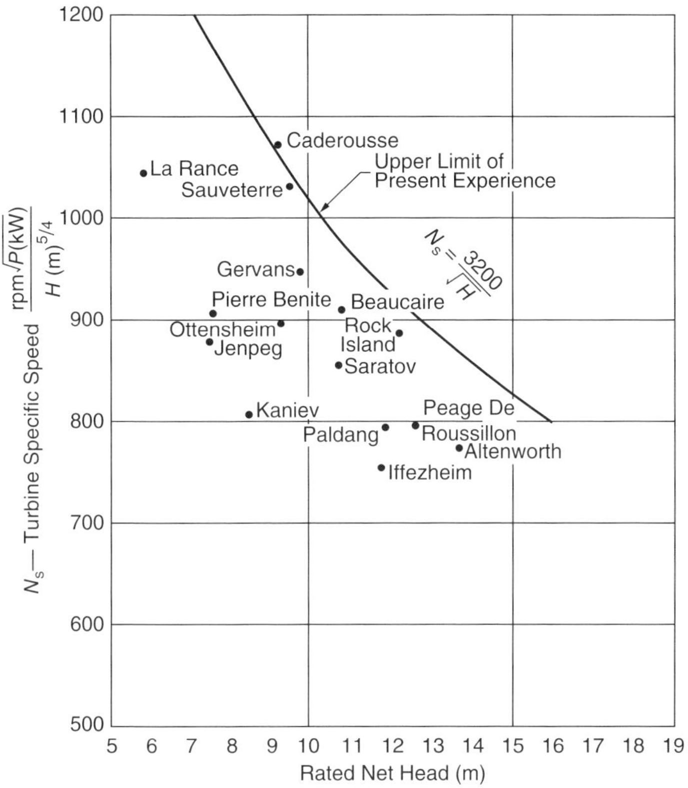

Figure 9.9 Experience curve of specific speeds for bulb turbines.

runner were reduced to a size that would develop 1 kW under one 1 m head. In the figures, the rated head is defined as the lowest head at which the turbine is capable of driving the generator at its rated output. Since most of the developments, which plot near the curve representing the upper limit of current experience, are run-of-river plants, their specific speeds correspond to outputs that are less than maximum, which suggests that tidal units might be selected with higher specific speeds since full-gate capacity for a tidal power development would occur only at infrequent times of very large tidal range. At the present time, the upper limit of specific speed would be about 1135. Benefits could be expected from adoption of tidal-electric turbine designs with higher specific speeds and unit powers. However, in pushing these characteristics to greater values care must be taken to ensure that turbine performance does not deteriorate and that the margin of safety against cavitation is maintained so as not to negate any benefits that might be achieved. Moreover, for higher specific speeds, the depth of setting of the turbine would likely have to increase and, depending on the site, the cost of deeper settings may be greater than

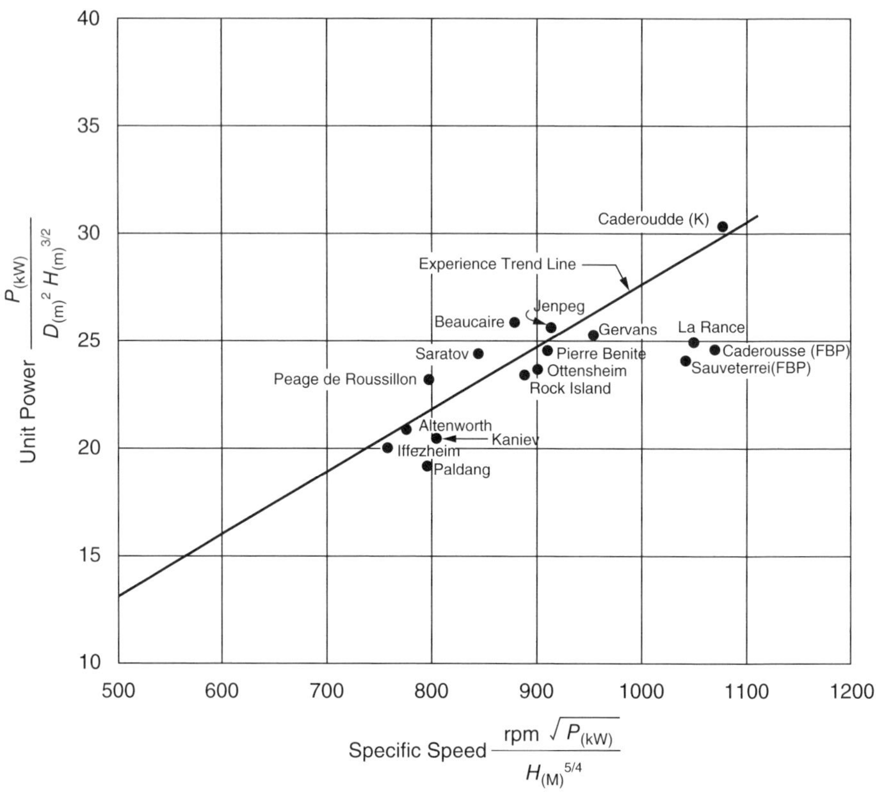

Figure 9.10 Relationship between unit power and specific speed for bulb turbines.

the value of the extra output. Such site-specific economic limits to increases in specific speed must be recognized.

9.3.5 Runner Diameter

Historically, there has been a trend toward a steady increase in machine sizes in an effort to reduce unit costs. Bulb units of 7.5 m diameter were placed in operation during 1977 at the Jenpeg Hydroelectric Station on the Nelson River in northern Manitoba. Manufacturers have suggested that low-head, axial-flow, horizontal shaft turbines with runner diameters as large as 9 or 10 m might be feasible. In this connection the relationships between power output, diameter and speed, as set out as (9.1), (9.2), and (9.3), should be kept in mind, that is, $P \sim D^2$, $n \sim 1/D$, and $n \sim \sqrt{H}$; where P = output, D = turbine runner diameter, H = rated head, and n = turbine speed. Since the output of a turbine of a given hydraulic design and specific speed varies as the square of the diameter, a 10 percent increase in diameter would result in a 21 percent increase in power output. On the other hand, however, the speed of the unit would be reduced by 10 percent. For the Fundy sites, a diameter of 7.5 m was adopted, for determining the degree of equipment installation and costs, since this dimension would not exceed the current state of the art. For the Severn Estuary studies, a diameter of 9 m was chosen as being a reasonable extension of existing technology and in the knowledge that the submergence required for the large diameter (to prevent cavitation) was available in the estuary without excavation.

9.4 GENERATORS

The design of the generator for a bulb unit is probably more critical than the turbine because of the restricted space inside the bulb. The common constraints are that the bulb diameter should not be more than 120 percent and not less than 80 percent of the runner diameter (Seoni, 1977). The former constraint is governed by the length of the conical section of the bulb and its effect on hydraulic performance; the latter constraint is imposed by electrical design. To meet these criteria, the inside stator diameter may range between 50 and 75 percent of that of a conventional vertical turbogenerator unit. To accommodate the large number of poles required for a slow speed unit in the restricted diameter, it is necessary to use "needle" poles.

It is important to keep in mind the relationship between the generator output and its dimensions and speed, expressed as

$$P = KD_r^2 L_n \tag{9.5}$$

where P = generator output (kVA)

K = compactness factor $\dfrac{\text{kVA}}{\text{m}^3 \times \text{r/min}}$

D_r = outside rotor diameter (m)

L = rotor length (m)

n = generator rotating speed (rpm)

For hydraulic and design reasons, the generator rotor diameter, D_r, must be considerably smaller than the turbine runner diameter. In order to compensate for this reduction, the generator length L and the compactness factor K must be increased. There are limits on increasing these values since, by doing so, temperature increases would automatically occur. In addition, with a high K value, the power factor has to be in the region of 0.97 to 0.99. Seoni (1977) has suggested that, based on past experience and realistic projections, 42 MW seemed to be the maximum rating at 50 to 60 rpm for units with a 7 m head.

Because of its smaller diameter, the rotor has an inherently low inertia, an undesirable feature in large power systems. As noted previously, tidal-electric units are particularly subject to rough operation, that is, pressure fluctuations from ocean waves, and are seldom in the optimum hydraulic operational range but usually at extreme conditions that automatically cause flow separations leading to power output fluctuations. Therefore, a large inertial value or Wr^2 (flywheel effect) would help to reduce the fluctuating loading of the critical connections of the poles and windings.

It should be noted also that accessibility is closely influenced by the compactness of a bulb generator. It is difficult to carry out extensive repairs, especially on the stator, without removing the generator. There are several ways of doing this, each of which requires additional cost and/or outage time. A maximum runaway speed of 2.5 to 3 times normal should not cause overstressing of either the rotor or shaft for a bulb unit.

A generator of about 50 MW capacity and 97 percent efficiency will require 1.5 MW of waste heat to be removed at full load. This may be done by air cooling or water cooling. For air cooling, fans at either or both ends of the generator force air through ducts in the stator and the rotor. The air is cooled in air/water heat exchangers. Because the power for cooling is inversely proportional to the square of the pressure, air-cooled bulb turbogenerators are usually pressurized to 2 or 3 bars absolute. Cooling at atmospheric pressure has the advantage that the bulb is normally accessible even when the unit is running. Water cooling of windings has obvious advantages but is considered to be the last choice, in view of the additional equipment and maintenance costs. This is all the more so with the large number of units required for a tidal development. Advantage can be taken of the cooling effect of water flowing around the bulb by provision of fins on the inside of the bulb; however, these measures are not sufficient for large machines, and separate air-to-water heat exchangers are required.

9.5 ELECTRICAL EQUIPMENT

The generator is obviously the key item of electrical equipment at the powerhouse. However, with bulb units, the generator is fabricated as an integral part of the turbogenerator unit. There are a number of electrical items to be considered in the final design of a tidal-electric plant, such as: (1) the excitation system design for which the static type is preferred with an excitation transformer and automatic voltage regulator for each group of four units; (2) the generation voltage that could provide economy at 13.8 kV by eliminating the intermediate stepup transformation; (3) possible integration of the switchboard into the powerhouse units with transmission to shore over only a few high-voltage cables; (4) use of one transformer for groups of four or eight turbogenerator units; (5) group transformers to reduce the number and utilize more effectively the transmission line breakers; (6) the transmission of the power to shore, probably by means of SF_6 gas insulated bus duct; and (7) the switching station and control eguipment.

The transmission of the block of power to shore would generally be over fairly short distances of up to 1.0 km or so and could be accomplished using continuous bus ducts located inside a concrete tunnel on or inside the access dike. Such an arrangement would be safe and easy for access and maintenance under all-weather conditions (Seoni, 1977).

9.6 TRANSMISSION

The incorporation of the output of a tidal-electric plant into existing power systems requires an analysis of transmission needs to transport the output to both local and remote markets.

As already noted, the output of single-basin, ebb generation schemes occur in the form of isolated slugs or irregular-shaped blocks of energy, conforming to the portion of the tidal range permitting generation. The transmission facilities necessary to incorporate the plant output into the power system network must, therefore, be capable of transmitting the total power and energy output to the markets. In addition, if the plant is designed for double-effect operation, these facilities must be able to receive energy from the systems for operation of the tidal power plant in the pumping mode.

The transmission analysis must be based on certain factors and assumptions: (1) designated market areas; (2) the magnitude of the tidal power transfers (MW); (3) the voltage level for ac or dc transmission; (4) establishment of power flow patterns on existing rights-of-way, if possible; and (5) whether the transmission is to be considered non-firm, so that transmission need not be added for reliability allowing for the loss of one circuit. A conceptual transmission configuration may be laid out for delivering various power magnitudes. The alternatives may range from 345 to 765 kV ac as well as high voltage direct current (HVDC), such as ±500 kV. For large power transfer levels, the HVDC will likely provide the least cost, even including the inversion and rectification costs. The transmission cost will depend on the terrain being traversed and upon the magnitude of the power transfers.

Detailed load flow and stability studies would be required to refine alternatives proposed in a prefeasibility study.

CHAPTER 10

OPTIMIZATION OF PLANT OUTPUT

For initial site screening the volume of a proposed barrage basin may be related to approximate annual energy estimates. However, accurate estimates of energy output depend upon a summation of the continuously varying production of energy since the operating heads change with the varying sea and basin levels, as well as upon the specific operating characteristics of the turbogenerating equipment. The major portion of this chapter is devoted primarily to analysis of single-effect, ebb generation for maximum energy production, without pumping. However, the same basic approach would apply, although somewhat more complex, to an analysis of the double-effect operating mode.

10.1 PRINCIPAL VARIABLES

Basically, the objectives of generation optimization are, first, to select the size and characteristics of turbogenerating machinery to suit site conditions and output requirements, and second, to determine the number of sluices and generating units that should be installed for maximum energy production to yield the lowest overall cost per unit of output for each location. The principal variables in the analysis, for any specific site, can be separated into two classes—those directly related to costs and those related to operating conditions and largely independent of cost, as follows:

Elements of Tidal-Electric Engineering. By Robert H. Clark

1. Cost-related variables
   a. Dike and sluiceway designs
   b. Turbogenerator design
   c. Numbers of sluices and turbines
2. Operating variables
   a. Time of commencing generation
   b. Time of concluding generation
   c. Timing of reversal and pumping cycles for other operating modes
   d. Runner blade and/or distributor gate settings during generation

The method outlined in this chapter for determining the optimum combination of sluice and turbine capacity requires: (1) mathematical modeling to find the maximum output attainable with a given set of sluice and turbine capacities; (2) using the models to determine outputs for a wide range of such capacities; and (3) comparing cost *versus* output over the whole range. A conceptual treament of these steps follows.

10.2 ENERGY PRODUCTION AND OPTIMIZATION

The variations in water level during a cycle of ebb generation are illustrated in Fig. 10.1. As described in Chapter 5, the generation phase commences at point A when the gross generating head is H_s and continues until point B when the head is equal to H_F. At this point, the turbine openings are closed off, the basin level remains constant, and the sea level continues rising. At point C, the levels become equal and the sluices are opened to allow the tidal basin to refill. If double regulated, the runner blades, as well as the

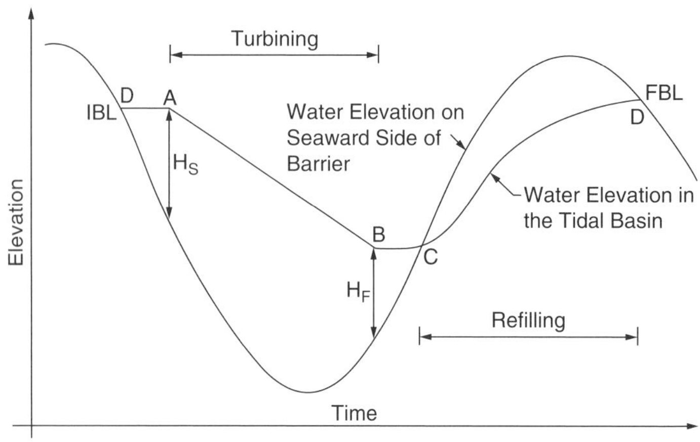

Figure 10.1 Ebb generation parameters for mathematical model.

distributor gates, can be adjusted to facilitate inflow to the basin through the turbine waterways. Basin filling continues until the levels again become equal at point D when the sluices are closed and the basin level held at its top elevation until the generation cycle commences again. For any selected turbine design, there will be a minimum value of H_s and H_F below which unit synchronization will be impractical. Moreover, to obtain maximum energy production, continuing operation to these minimum values is not necessarily required. There are, in fact, an infinite number of operational timings for each level of installation, only one of which will yield maximum energy production.

In simple terms, it may be imagined that if a small number of turbines were used for ebb generation, only a small amount of water would be released from the basin and the basin water level would remain relatively high. As illustrated by Fig. 10.2, if the number of turbines were increased, more water would be released, in a given time and the basin level would fall off more rapidly.

The optimum operating condition can be effectively determined by a computerized mathematical model. The objective of the model would be to produce estimates of energy production at a selected site for several combinations of turbine numbers and sluiceway capacities, and, by obtaining a cost estimate of each combination, it would be possible to select the installation that produces the most economical energy at a site, that is, the optimization of the scale of development. A characteristic site optimization curve is illustrated in Fig. 10.3.

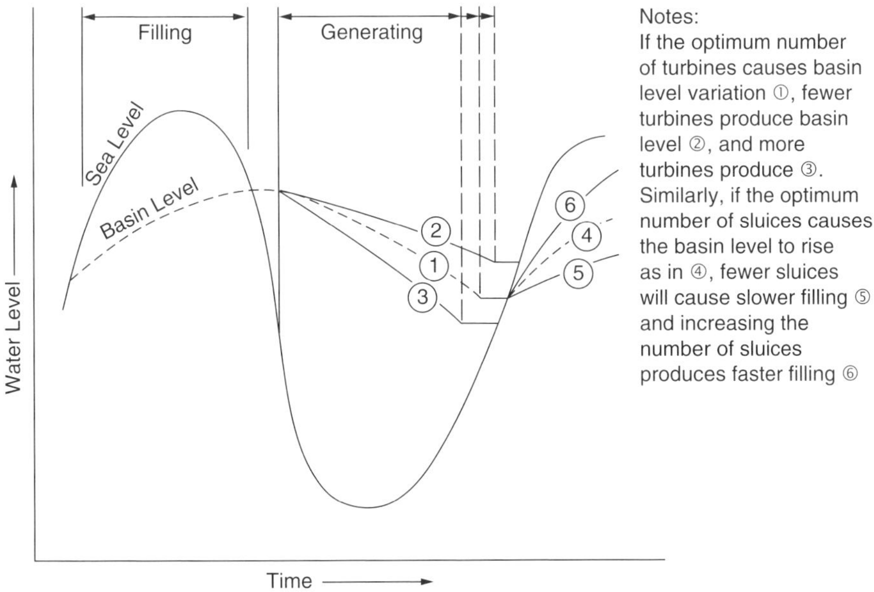

Figure 10.2 Effect of varying numbers of sluices and turbines.

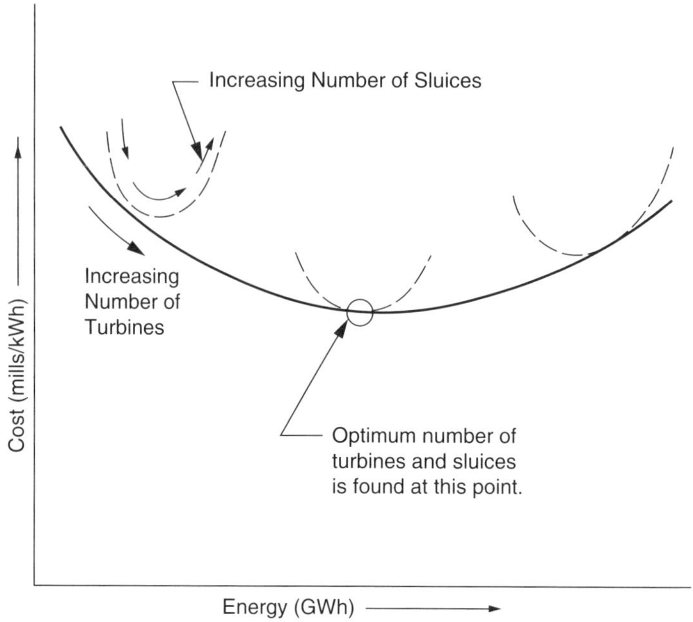

Figure 10.3 Characteristic optimization curve.

10.3 SIMULATION OF OPERATION

In order to simulate with reasonable accuracy the operation of a tidal-electric plant, the mathematical model should account for the following principal features:

1. The tidal basin shape in terms of related water elevations and surface areas from which volume changes can be deduced
2. A correction for the "backwater effect" in the tidal basin during plant operation in either direction
3. The shape of a typical tidal cycle expressed as a time series of water surface elevations
4. The use of either fixed-blade or variable-pitch runners accounting for efficiency variations
5. The operation of the sluices and of the turbines as orifices, with discharge expressed as a function of head
6. Variations in startup and finishing conditions with respect to turbining and sluicing

When such characteristics are included in the model, appropriate variations in tidal amplitude, turbine characteristics, and numbers of turbines and sluices can be introduced in each run of the model program to facilitate the estimation of optimum output for comparison with the estimated project cost. One consideration in development and use of simulation and optimization models is that programming cost increases and transparency and robustness may be adversely affected by increasing detail and sophistication of such programs. Fairly sophisticated models can be developed to allow greater freedom for the determination of optimum conditions, irrespective of operation mode. However, the degree of sophistication of the model used should be commensurate with the objective, purposes, and scope of the investigation.

10.4 DEVELOPMENT OF MODELS

The models used for the Bay of Fundy (Canada) reassessment study (Bay of Fundy Tidal Power Review Board and Management Committee, 1977) are discussed here only for the purpose of illustrating the basic mathematics and logic of the simulation and optimization. The development of these optimization models is presented in Appendix A for information. Other models may be more appropriate under tidal characteristics different from those of the Atlantic Ocean, particularly where the diurnal inequality is more pronounced. However, the objective would be the same: to estimate the energy generated at a tidal-electric plant of given characteristics by simulating its operation under the tidal characteristics at the site.

10.4.1 Single-Tide Optimization Model

The single-tide optimization model (STOM) is based on the *Swales–Wilson model* and incorporated many computationally efficient improvements. The model is relatively simple and cost efficient for optimization purposes but subject to certain limitations.

Its optimization logic is based upon the operation of a tidal-electric plant during an infinite series of identical tides. Under such a regime, the operation of the plant would be identical from one cycle to the next when operating against a constant objective function, that is, the objective function is not time dependent and the tidal basin levels at the beginning and end of the cycle are equal. Hence, a single tide cycle may be considered and the operation during that cycle optimized with the constraint that the basin level at the beginning of the cycle is identical to the level at the end. By examining a number of single tidal cycles of different ranges and optimizing the output for each separately, a close estimate of annual energy production can be derived using a tidal range histogram for the site. STOM also provides for internal storage of the step-by-step operational details that may be used to prepare hour-by-hour energy output estimates for use in studies of system integration. The model is very fast computationally. The model developed for the Mersey Barrage energy studies is similarly based (Potts and Wilson, 1992).

10.4.2 One-Cycle Dynamic Programming Model

Since in reality consecutive tides are not equal in amplitude, and in order to account for more complex operating objectives than maximum energy production such as, for example, the variation in the value of energy to the electrical system, a more general model, *one-cycle dynamic programming model* (OCDPM), was developed for the Fundy tidal power reassessment studies to simulate operation during the natural series of tides so as to define an optimum operation.

Dynamic programming models produce an optimum solution over a finite series of tides. Actual operation of a tidal plant must consider an *infinite* number of tides. However, use can be made of the quasi-periodicity of natural tides and the fact that an operation during a certain tide influences conditions of operation only during the next 2 or 3 cycles. Similar tide elevations are, as a close approximation, reproduced after a lunar month of 54 tides, and a computation over about 60 tides is therefore sufficient so long as the last few cycles are subsequently discarded to account for the fact that, for those cycles, optimization does not properly account for the future operation.

10.4.3 Operation for Maximum Energy

Since the OCDPM is significantly more complex to use than STOM, the former can be used to adjust or provide correction factors to the results from STOM. It is worthwhile to note that, since the Bay of Fundy tides are quite regular with only a small diurnal inequality, STOM produced energy estimates over a sequence of Fundy tides that were well within 1 percent of the total predicted on the basis of OCDPM. On the other hand, OCDPM can be used to optimize energy production according to an energy value function that varies with the time of day.

Single-effect operation (without pumping) for maximum energy requires: (1) the maintenance of the tidal basin level as high as possible at the beginning of the tidal cycle; (2) holding that level until the appropriate head becomes available for turbine operation; (3) turbining until the head approaches the minimum operating head for the turbines; and (4) refilling as rapidly as possible to bring the reservoir back to the highest level. This operation is illustrated in Fig. 10.1.

The single-effect optimization problem using STOM reduces to one of manipulating the starting and finishing times of the generating period while maintaining the same basin level at the beginning and end of the tidal cycle. The model assumes, in the first instance, that the finish of the turbining period occurs at the minimum operating head of the turbines. At three initial basin levels, y_1 the starting head, H_s of the turbining period is examined until the head for maximum energy production is found. These computations are made without reference to the refilling period. A function relating the starting head for maximum energy, $H_{sE_{\max}}$, to the initial reservoir level is derived as follows:

$$H_{sE_{\max}} = A_0 + A_1(y_1) + A_2(y_1)^2 \tag{10.1}$$

where A_0, A_1, and A_2 are computed coefficients. Using this relationsnip, y_1 is adjusted until the level y_2 at the end of the refilling period is equal to y_1. The energy production from this analysis is very close to the maximun energy production from this single tide cycle, at least for those locations where the diurnal inequality is small.

Double-effect operation to maximize energy production requires rapid drawdown and refilling of the reservoir at the end of each generating period to obtain maximum heads for turbining during the subsequent half cycle. Consequently, although energy production continues until the minimum operating head of the turbines is reached, the sluices are opened before that head is reached. This operation is illustrated in Fig. 10.4. Thus, with two-way or double-effect generation, the turbines would operate somewhat less efficiently at any given head than for single-effect or one-way generation since optimization for the latter condition must be compromised to obtain an optimum design for double-effect operation.

The double-effect optimization problem using STOM relates to the selection of the time (or head) when the sluices should be operated in addition to the time (or head) when turbining should start in order to maximize energy production while keeping the initial and final tidal basin levels equal. The search for the optimum becomes somewhat more complicated than that for single-effect operation because a change of one variable disturbs all others. A solution to the problem can be obtained by fixing the operating head for the sluices, H_{slu}, in each half cycle and to adjust simultaneously the starting heads, H_s, in each half cycle to obtain maximum energy production. Then, by

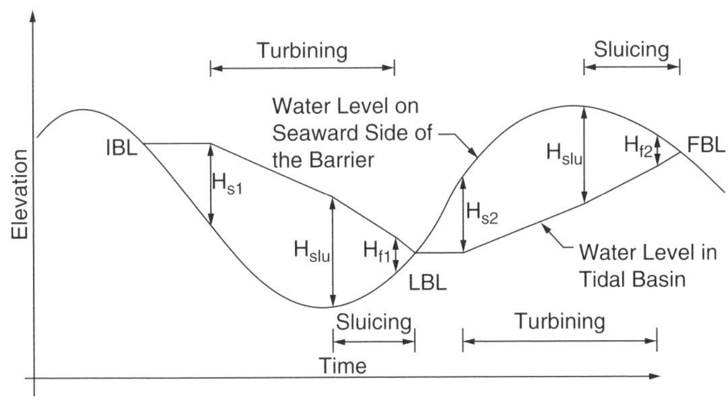

Figure 10.4 Double-effect operation parameters for single-tide optimization model.

maintaining both values of H_s constant, H_{slu} can be adjusted in each half cycle to obtain a higher production. Finally, H_s in each half cycle is adjusted while maintaining H_s constant in the other half cycle and holding H_{slu} constant in both half cycles. This approach gave consistent results for the Fundy tidal conditions and also gave results in close agreement with tne OCDPM for double effect, which is based on a more rigorous approach.

It should be noted that the energy function for a tidal cycle subjected to double-effect operation is not unimodal as in the case of single-effect operation. There will be two local maxima—one in which the operation tends to single effect and a second in which the operation stabilizes in the double-effect mode to produce an energy value in each half cycle that would be more nearly equal in those locations where the diurnal inequality is low. The first maximum would be found for low and medium tide ranges and the second for high tides.

10.4.4 Complex Modes of Operation

Other modes of operation such as single effect with pumping, double-effect operation with and without pumping, using variable-pitch turbine blades, and the like may be optimized with the one-cycle dynamic programming model. Suboptimization routines can be developed (Bay of Fundy Management Committee, 1977) for such modes of operation.

10.5 PLANT OPTIMIZATION

Because of the number and complex interrelationships of the variables involved (refer to Fig. 10.5) in satisfying an objective function, such as minimizing the cost per unit of output from a tidal-electric plant, an analytical solution to site optimization is remote if the problem is to be treated with rigor. Therefore, the approach to optimization must be through an iterative process or search technique that should be made as cost-effective as possible.

Using a model, such as STOM, the best operating strategy and the output can be computed for one complete tidal cycle at one level of installed capacity, one level of sluice capacity and one set of turbine characteristics. This procedure would be repeated for other values of installed sluice capacity to provide data for one of the sloping lines in Fig. 10.6. Data for other values of sluice capacity are similarly obtained. A capital cost estimate for each combination would be prepared according to a costing formula developed in the plant design studies. From this capital cost, an estimated annual cost would be computed to cover amortization, interim replacement, operation and maintenance.

For each combination of turbines and sluices, a minimum of three tidal ranges would be selected to encompass the extremes of the annual tidal

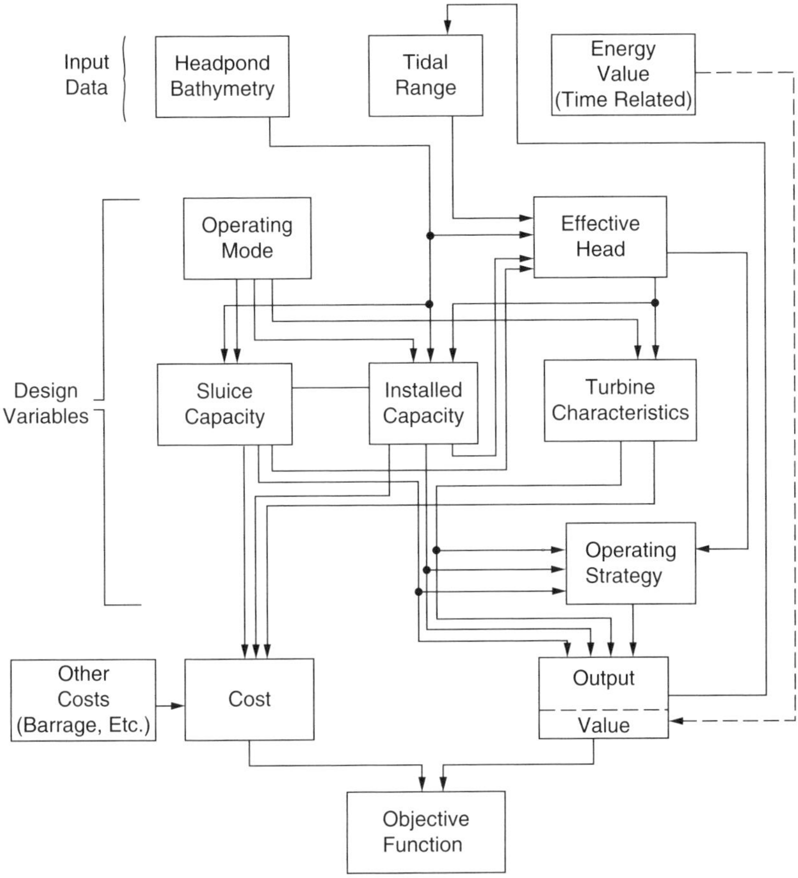

Figure 10.5 Main interrelationships of variables for optimizing a tidal-electric plant.

histogram. The optimum energy output for each range could then be calculated using the STOM or a suitable model, and a relationship between energy production and tidal range developed. The total annual output would then be determined by reference to this relationship and the histogram. The total annual output and annual cost will yield the mean annual unit cost of energy. Curves may then be plotted for each level of installed capacity with unit costs for varying levels of sluice capacity against the computed annual output, and the optimum point of the curve located. Repetition of this sequence for other levels of installed capacity would provide further optimum points. A curve indicating the best combination of sluices and turbines for a particular site can be drawn, as illustrated by the example in Fig. 10.7. Adjustments to the tidal range would also have to be made for each combination of installation to account for the effect, if any, of barrage operation on tidal amplitude.

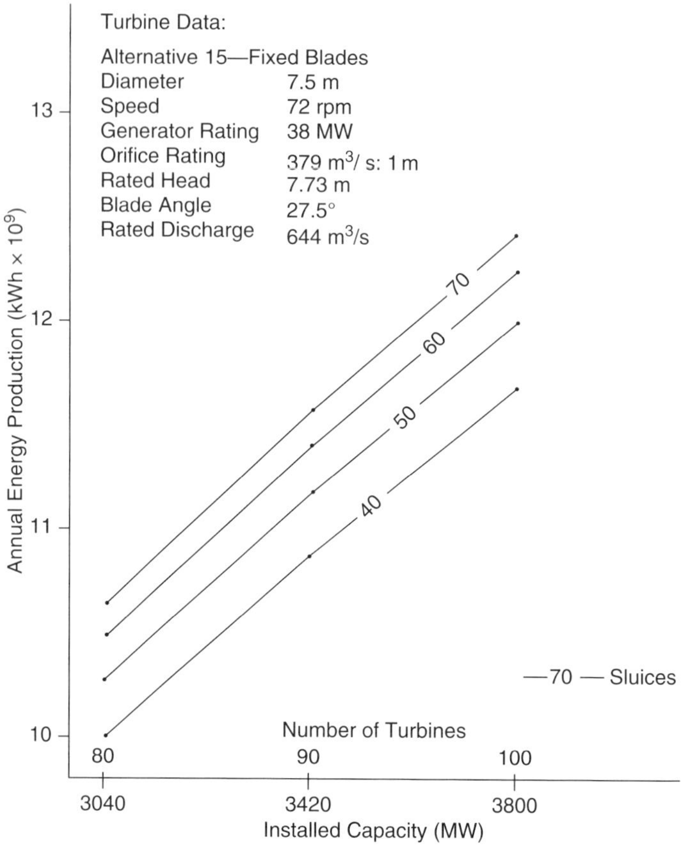

Figure 10.6 Annual energy production versus installation levels for ebb generation at Site A8, Bay of Fundy with various number of sluices.

In determining the flows used for generation, turbine performance curves, such as those illustrated in Fig. 9.8, would be utilized. As pointed out in Chapter 9, such unitized characteristic curves must reflect all intake and draft tube losses. Head losses through the intakes and draft tubes are not necessarily accounted for in such unitized characteristic curves. For fixed-blade installations, the units would be operated along the maximum power and limiting power curves at all stages of the tide, resulting in a deterministic relationship between head and discharge. Although this assumption would introduce a slight error in the assessment of maximum production, it would amount to a loss of only about one-half of one percent of the annual energy production in comparison with the true optimum; an adjustment to the annual energy production could, therefore, be made, if appropriate. Operation of variable-pitch blade turbines for maximum energy production can be simulated in accordance with the more complex pathway through the turbine hill chart

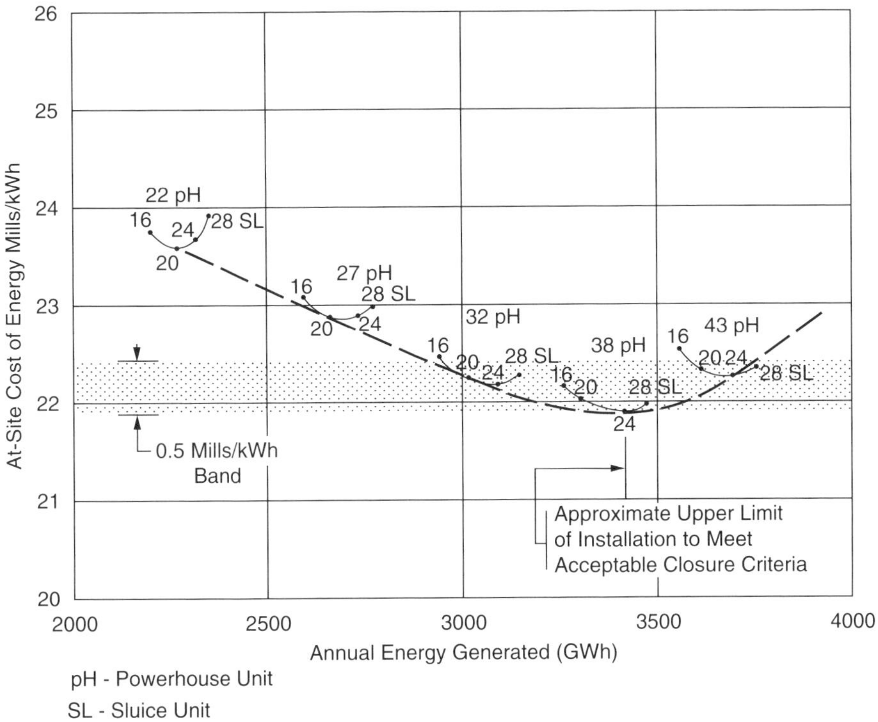

Figure 10.7 Single-effect optimization for Site A8, Bay of Fundy.

determined by introducing the calculus of variations to define an optimal (maximum energy or benefit) relationship between discharge, reservoir level, and time during the generating period (Fig. 10.4).

For basin refilling, all turbine openings should be assumed to perform as orifices to augment the sluicing capacity. The sluices, if fully submerged venturis, would have a discharge coefficient of about 1.5, based on the results of considerable hydraulic model testing. The discharge coefficients for both fixed and variable-pitch turbine runners when operating as orifices are not known precisely in the absence of available model test data but would probably be of this order. In any event, the discharge through the sluiceways would be several times larger than that through the *turbines as orifices* so that small variations in the discharge coefficient for the latter would be acceptable for prefeasibility studies.

Additional studies could be undertaken to evaluate alternative turbine runner diameters, fixed-blade angles, and alternative generator ratings for a given turbine. In the Bay of Fundy reassessment studies, it was found that the cost of energy was relatively insensitive to these alternatives. Studies of single- and double-effect operation, with and without pumping, to establish the relative costs and energy production of such schemes and their value to the system should be assessed (refer also to Chapter 5).

With reference to Fig. 10.7, it is interesting to note that the at-site optimization curve tends to be relatively flat across a broad range of possible energy outputs. Optimization curves for the Severn Estuary (Great Britain) and Cook Inlet (Alaska, U.S.) studies showed a similar characteristic flatness.

CHAPTER 11

INTEGRATION OF OUTPUT WITH ELECTRIC UTILITY SYSTEMS

11.1 INTRODUCTION

Since the lunar period is slightly longer than the solar day by about 50 min of solar time, the times of high and low tidal levels, occurring in phase with the lunar cycle, shift slightly forward in solar time each day. This phase difference changes the maximum quantities of tidal energy that can be generated at any particular time each day. In addition, tidal energy generated by a single-effect, tidal-electric plant occurs for only two periods of several hours duration each day (Fig. 11.1). A sequence of such tidal power pulses over a period is shown in Fig. 11.2, taking into account this quasi-sinusoidal variation in the magnitude of high tides during a month since spring tides occur at bimonthly intervals (refer to Chapter 2). Therefore, the absorption of tidal energy by power systems, which have varying load demands during daily, weekly, and seasonal cycles, poses certain problems. However, a very positive plus to a tidal plant's output is the predictability of time of occurrence, magnitude, and duration of the tide. Earlier studies of tidal power generally ignored the operational feasibility of absorbing "raw" tidal energy, that is, the energy as it is generated, by a power system and also the modification it would produce in the generation expansion plan of the affected system.

A relevant question that must be answered is: How much of the energy produced by a tidal-electric scheme can be absorbed directly as raw energy? The rate of absorption at any time is limited by the system factors discussed

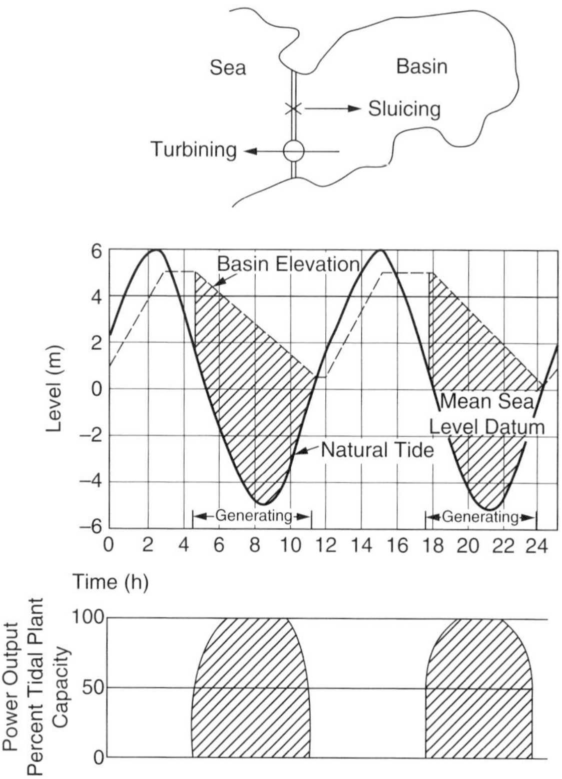

Figure 11.1 Single-basin, single-effect scheme and its output.

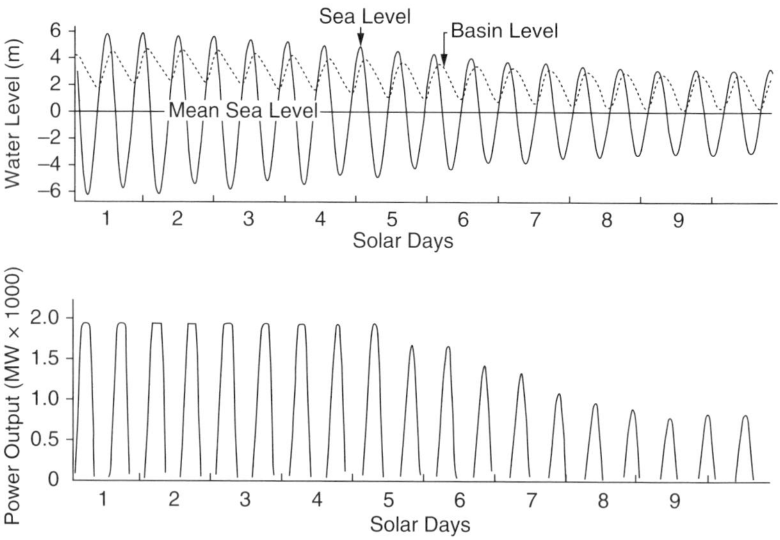

Figure 11.2 Typical tidal range and energy output during a spring-to-neap cycle for a single-effect plant in the Bay of Fundy.

here, and may be further limited by the amount of tidal energy available at the time. Power systems normally adjusted supply to demand by regulating the production from hydroplants, by varying the output of large thermal plants within certain limits, by adding power from gas turbines during peak load hours, and through the use of retiming facilities, such as pumped-storage plants. Assuming that there is a mix of generating sources in a utility system, Fig. 11.3 illustrates the procedure for allocating the capacity of these sources on the load curve to meet the system load demand. The *must-run* plants would be operated to supply the energy at the base of the load–duration curve, but allocation of the remainder of the load would be based on incremental costs of energy. Thus, the next step up on the load curve would be supplied by the most efficient thermal stations in the system, and they would be providing a large block of energy with the less efficient (older) thermal stations supplying energy higher up on the load–duration curve to minimize the overall thermal operating costs.

To supply the peak, that is, the portion of the load forming the top of the load–duration curve, there may be several choices, such as gas turbines, peaking hydropower stations, and pumped-storage plants. It should be noted that the positioning of the hydroplant output near the top of the load–duration curve means that *capacity* and not *energy* is the dominant factor. It should also be kept in mind that plant availabilities are not constant throughout the year, and that the level of available capacity for each type of plant will depend on

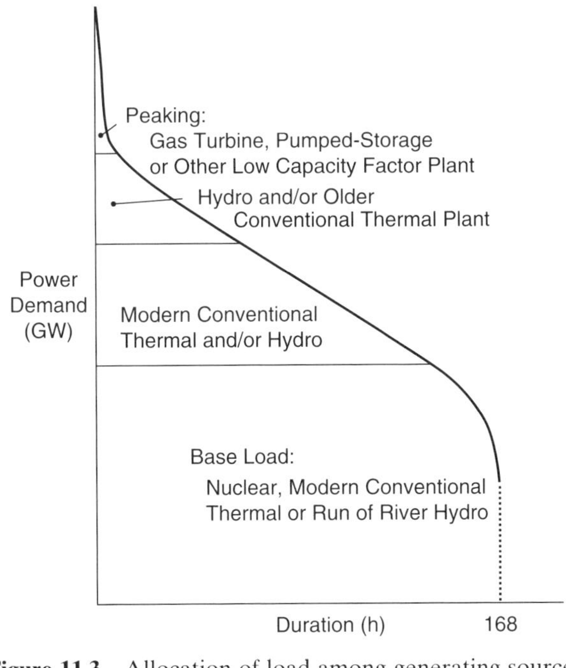

Figure 11.3 Allocation of load among generating sources.

the scenario being considered. One may now ask: How much raw tidal energy can be absorbed by making full use of the already existing market supply with the lowest marginal costs? And the bottom line of the issues: What is the economic value of the tidal energy absorbed?

Currently, there is a trend in a number of countries to the restructuring of electricity markets to permit open access to utility transmission systems and to create supply options. As a result, economic dispatch of energy would not be constrained to a utility's own generation. For example, in the United Kingdom, the Central Electricity Generating Board has been dismantled. Transmission is operated by the National Grid and there are two dominant suppliers and many minor suppliers. In the United States, the Energy Policy Act of 1992 removed many of the geographical and regulatory impediments to the wholesale wheeling of electrical energy and has opened the way to compulsory transmission access. In Canada, the Province of Alberta has created a competitive market whereby the utilities and nonutility generators (NUGs) sell into a pool and the utilities purchase from it. The implications of this "free-market" approach with regard to the exploitation of tidal energy places greater emphasis on the question of retiming, but makes it less easy to resolve since it would tend to make marginal costs the criteria of saleability, whereas utility cost includes the fixed cost of its own generation.

However, deregulation of, and competition in, the electricity supply industry poses several problems, such as: Who looks after the long term since competitive power markets tend to be very short term focused? Who looks after research and development and the demonstration of new technologies? How will governments meet their international commitments to reduce national CO_2 emissions if they can no longer dictate fuel choice and require utilities to subsidize customers' end-use efficiency through demand-side management programs?

11.2 ABSORPTION OF RAW TIDAL ENERGY

11.2.1 Static Absorption Capability

Tidal energy absorbed by a power system has the effect of displacing or reducing a like amount of thermal energy generated within that system. The value of energy saving associated with tidal energy can thus be measured in terms of the amount and cost of thermal energy displaced. It may also be possible to reduce the total installed capacity of nontidal generation when a tidal-electric plant, even an ebb flow generating one, is brought into the system. This may also result in slight capacity savings that could be credited to the tidal-electric scheme.

Will there be any tidal energy "surplus" after all the tidal energy that can be absorbed by a system has been absorbed? What are the characteristics of this surplus? What can be done with it? It might be used for retiming by means of a pumped-storage facility, although it would be logical and utility practice to use, for this purpose, the cheapest energy available after meeting current loads. More complex tidal schemes could be built to gain more control over the output although, as pointed out in Chapter 5, this alternative usually means a substantial reduction in the total energy output from the site as well as a substantial increase in the cost of that output. On the other hand, the surplus power could be exported to contiguous systems over existing or specially built transmission lines.

The static absorption capability of a system may be defined as its ability to reduce or "back-off" generation in order to accept the periodic tidal energy outputs illustrated in Fig. 11.2, or portions thereof. Generation can be backed-off to a level that is referred to as the "must-run" level. The types of must-run generation would be the following:

1. Must-run hydro Technical and/or operating constraints prevent hydro generation at some plants from being reduced to zero, for example, to maintain a minimum river flow or to avoid spillage; this portion of the hydro capacity is must-run generation.
2. Must-run nuclear generation Nuclear plants are not designed to follow rapid load changes, and, even if they could cycle, the fuel saving obtained by tidal substitution should the variable cost of nuclear be higher than the average cost of tidal, would not be significant; a specified fraction of the nuclear capacity in the system constitutes a component of must-run generation.
3. Minimum oil- or coal-fired, must-run generation Shutting down completely fossil-fueled generators twice a day for the benefit of tidal absorption would incur significant startup costs and increased O&M costs; some fraction of load in excess of the nuclear and hydro must-run is considered as fossil must-run, which corresponds to the minimum loading of the fossil units and depends on the operating characteristics of the plant or plants committed.

In summary, then, the amount that can be absorbed is a function of: (1) the system size and load characteristics; (2) the rate at which the output from other generating units can be backed off and displaced by tidal generation; and (3) also the rate at which these units can be ramped up again when generation from the tidal plant has terminated. Figure 11.4 shows a typical annual load–duration curve with the duration of the must-run generation plotted on an annual basis. It is only the energy demand represented by the area above the must-run line that is available for the absorption of raw tidal energy. For the analysis, the actual duration curve of must-run generation is replaced by the average must-run line, which yields the same must-run generation area. The area above the average must-run line is subdivided into equal segments, ΔL, of system load, L. The annual load–duration curve of a tidal plant is also subdivided into horizontal segments of height ΔG such that

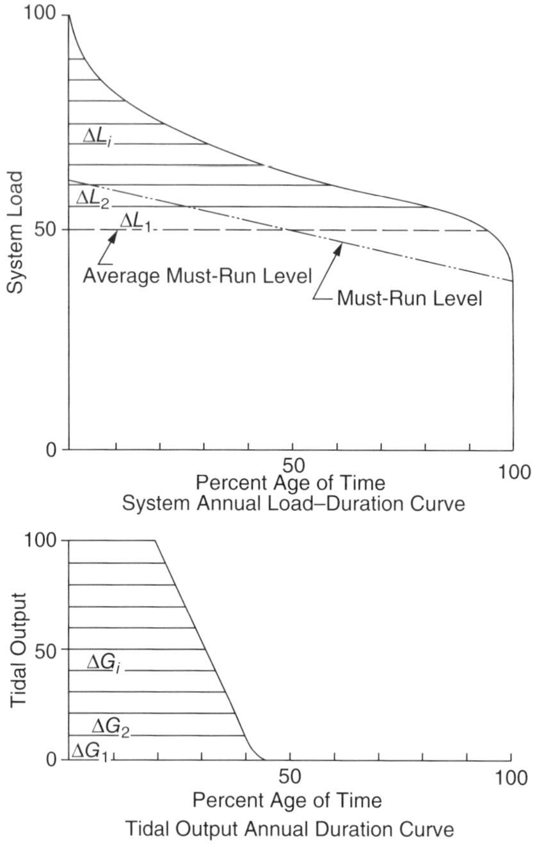

Figure 11.4 Typical relation of "must-run" generation and tidal energy output with respect to a system annual load–duration curve.

$$\Delta G_i = \Delta L_i \tag{11.1}$$

and the percent duration of time for which ΔG_i and ΔL_i exist is t_{G_i} and t_{L_i}, respectively. By using the method of convolution of two cumulative probability or distribution functions, that is, of the system load–duration and tidal output duration curves, the raw tidal energy that a system is able to absorb can be determined. Since the 705 tidal impulses per year are uniformly distributed and the length of a spring-to-neap tidal cycle is small compared with the simulation period of one year, the length of time, t_{a_i}, during which the tidal output, ΔG_i, can be absorbed into the system load, ΔL_i, may be calculated as

$$t_{a_i} = \frac{t_{L_i} - t_{G_i}}{100} \tag{11.2}$$

Equation (11.2) enables the construction of a duration curve of absorbed raw tidal energy from the available annual load duration curve and the tidal output duration curve. The area under the duration curve of absorbed tidal energy gives the amount of energy directly absorbed in the system. This total absorbed energy, as a fraction of the generated energy, is given by

$$e_a = \sum_{i=1}^{n} \frac{t_{a_i}}{t_{G_i}} \tag{11.3}$$

where e_a is the amount of tidal energy absorbed directly by the system. The error in using this simplified method, which has been developed and presented by Furst and Sud (1977), is small.

11.2.2 Dynamic Absorption Capability

In view of the limitations on static absorption capability, tidal generation would primarily displace coal, oil, and gas-fired generation. These plants would have to be capable of being "backed-off" and accepting loads at rates dictated by the superposition of the steepness of the tidal pulse and the coincident load changes in the system. The resultant load-following capability required from the cycling plant is

$$\frac{dF}{dt} = \frac{dT}{dt} \pm \frac{dL}{dt} \text{MW/min} \tag{11.4}$$

where, dT is the steepness of the front or tail of the tidal pulse in MW/min, and dL/dt is the rate of change in the system load in megawatts/minute.

Figure 11.5 illustrates a load curve modified by the absorption of a tidal energy pulse that would now have to be followed by "cycling" generation. The maximum rates of change of output from tidal schemes of three hypothetical capacities would be about as follows:

Capacity, MW	1000	3000	7000
Increasing output, MW/min	50	100	200
Decreasing output, MW/min	30	40–60	120–150

The output gradients of a tidal plant can, of course, be reduced substantially at a very small cost in energy. For example, effective gradients can be reduced to half these maximum gradients by sacrificing less than 0.5 percent of the available energy.

Thermal plants of recent design probably have a load-following capability of not more than 5 percent of rated output per minute above a minimum load level. Older thermal plants will not be able to follow load as well as new units, and system control limitations will cause a group of jointly controlled plants to have a load-following capability less than that of one individual plant. For example, if it were assumed that the average load-following capability of plants assigned to cycling duty were 3 percent of rated capacity per minute,

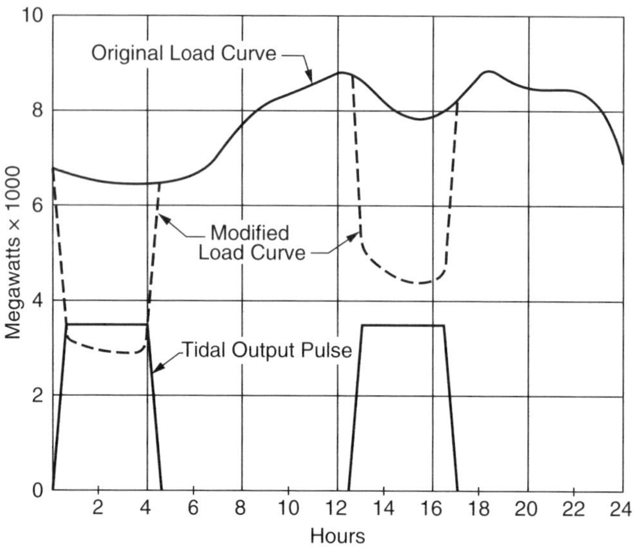

Figure 11.5 Typical daily load curve modified by tidal energy absorption.

the Nova Scotia–New Brunswick systems, which have a combined capacity of about 6000 MW, could manage the absorption of the raw as delivered output of a 1500-MW tidal plant in the Bay of Fundy without serious problems. The Nova Scotia–New Brunswick systems also include hydro storage, which could also absorb tidal energy. It would be realistic to assume, however, that by the end of this decade, the average load-following capability of plants assigned to cycling duty would be about 10 percent of rated capacity per minute above minimum loading levels. As already indicated, dynamic limitations may result in either a very small loss of raw tidal energy or nonoptimum absorption, but the overall effect would be too small to warrant consideration of dynamic limits during prefeasibility studies.

11.2.3 Summary

Raw tidal energy absorption capability of a system varies with changes in system load and must-run generation change, but it is essentially the difference between system load and must-run generation at any given time. The changes are sufficiently slow to allow an hourly simulation with sufficient accuracy for prefeasibility studies. However, a more detailed analysis of the dynamic problems associated with the output gradient of tidal plants would require system simulation in much smaller time steps than one hour. The conceptually simplest approach for calculating annual absorption of raw tidal energy is to develop a chronological sequence of 8760 hourly load must-run and tidal output data, and then to compute the amount of raw tidal energy absorbable by the simple process of subtraction. However, feasibility studies require the consideration of numerous alternative tidal schemes and market configurations, which makes this approach rather expensive and time consuming. A semiprobabilistic approach developed by Furst and Sud (1977) avoids hourly simulation but yields excellent agreement with the detailed hourly system simulation.

The dominant parameters affecting the absorption capability are the ratio of the maximum tidal output to peak system load, the must-run level, and the annual system load factor. The must-run level is governed by both static and dynamic considerations. The dynamic constraint, arising from the limited load following capability of a thermal plant, is only significant in the case of tidal plants that are very large with relation to the system peak. The lower the must-run level and the higher the annual load factor, the higher is the raw tidal absorption capability for the same tidal plant/system peak ratio.

11.3 ENHANCING RAW TIDAL ENERGY OUTPUT

As outlined in Chapter 5, the single-effect, single-basin, tidal-electric plant is an intermittent producer of energy, the periods and nature of production depending upon the tidal cycles. Various modes of operation of a tidal development are outlined, including double-effect operation, pumping, the use of a double-basin and even linked-basin schemes, in order to provide a more continuous output and confer a *dependable* or *firm* capacity benefit to the development. However, the more extensive the civil works required to accomplish this end, as well as the more sophisticated the generating equipment, results in a considerably higher unit cost of energy since substantial energy must be sacrificed to adjust the timing of the output.

There are four basic options for energy storage, external to the development itself, to regulate a tidal plant's contribution to the electric system in which it will operate. These options can manipulate the raw output and make it more compatible with the daily power demand pattern and provide firm capacity as well as energy. These basic options are: (1) coordinated operation with river hydro, (2) pumped storage, (3) compressed-air energy storage, and (4) auxiliary capacity.

11.3.1 Coordinated Operation with River Hydro

If a tidal power development is contemplated for a system that has a large hydro component with adequate water storage facilities, then an economically attractive solution of retiming the raw tidal energy is available by operating the tidal plant in conjunction with the river hydro plants. Such an option would require available water storage equivalent to about one cycle of tidal energy and that the hydro component would have sufficient spare generating capacity and also that its normal output would equal or exceed the peak output of the tidal plant. This option has the advantage that it would entail no significant loss of energy, and the only additional costs would be those relating to providing the required spare capacity, if it did not already exist.

A variation of this option could involve the utilization of very large storage developments, which may exist in a contiguous and interconnected power system. However, this alternative could involve fairly long transmission distances and possibly higher power losses, depending upon the energy flow routes during the drawdown and repayment periods.

11.3.2 Pumped Storage

The energy production of a tidal plant, occurring at times when it would have a relatively low value to the system, could be used as input to a pumped-storage facility. The transmission requirements to interconnect the tidal plant, the pumped-storage plant, and the contiguous power systems, would depend on the installed pumped-storage capacity and upon the manner in which the operation of this facility is integrated with that of other system facilities (Warnock, 1971).

Pumped-storage capacity requires, in principle, close proximity with water resources of sufficient volume and storage reservoirs of suitable capacity, sufficiently close to each other and at an adequate difference in elevation so that economic pumped-storage facilities can be constructed. Based on these principal requirements, pumped-storage capacity may be found (1) on or adjacent to river systems and (2) adjacent to the sea, using seawater. The storage sites should be close to the tidal plant to avoid transmission losses, geologically adequate, and provide suitable conditions for the power plant and interconnecting waterway between upper and lower reservoir. If seawater is to be used, the storage site must be free from seepage of saline water, which could contaminate freshwater resources and arable land. Underground reservoirs overcome this potential problem.

The basic function of a pumped-storage development, as ordinarily conceived, is the storage of low-quality, off-peak energy for subsequent reuse as high-quality, peak energy. In most predominantly thermal power systems, as indeed in all power systems, there is considerable daily variation in the load demand—the minimum demand may often be 30 percent to 40 percent of the maximum daily peak load. In most power systems, it is necessary to meet such load demands with conventional fossil-fuel-fired generation or nuclear generation. Neither of these forms of generation is well adapted to carrying extreme load variations because of costly penalties associated with such operation, such as fuel and reliability. Better system performance and lower costs generally result with the addition of pumped-storage capacity to the system power resources. This permits using excess electrical energy produced during the night, low-demand hours to pump water from the lower pool to the higher pool, storing the electrical energy as potential energy for later use during peak system demand by reconverting it into electrical energy by releasing it through generating units, usually those used for the original pumping operation operating in reverse. The pumped-storage arrangement may be such that the water is recirculated so that only seepage, leakage, and evaporation losses need be replaced. Further information on types of electrical and mechanical machinery, the role and importance of pumped storage in power systems and in water resources development, and effects on the environment may be found in the Proceedings of the International Conference on Pumped Storage Development and Its Environmental Effects (Karadi et al. 1971) and in the Report of the International Commission on Large Dams (1980). Thomann (1978) discusses environmental aspects in pumped-storage planning.

Pumped-storage capacity used in conjunction with a tidal-electric plant would be operated in a slightly different manner to that required in a peaking capacity, its conventional role. As already noted, either night-time or off-peak energy developed by the tidal plant is converted into potential energy, stored, and reconverted into electrical energy, as needed, to meet system operating contingencies as well as peak demand. Its most important advantage is, therefore, the ability to convert low-quality components of generation into higher-quality production so as to maximize the return. A pumped-storage plant also has the ability to pick-up system load quickly when functioning as system reserve or operating at less than capability—an ability very much superior to that of any form of generation other than hydroelectric. However, it should be noted that, even with the high efficiencies of the pumping and turbining operations, the pumped-storage arrangement will only deliver 75 to 80 percent of the input tidal energy at most.

It may be that there are no topographically suitable sites for pumped-storage capacity within acceptable transmission distances to the load centers. Even when suitable surface sites are available, they are often the subject of considerable opposition from the environmental standpoint. Underground pumped hydro employs the same basic principles as conventional, surface-sited pumped storage except that the head is created between an upper reservoir at ground surface and a lower reservoir, located in cavities excavated in rock at depth (Warnock and Smith, 1984). The underground storage space may be chambers excavated in sound rock or existing caverns or openings, such as an abandoned mine.

11.3.3 Compressed-Air Energy Storage

Compressed-air energy storage (CAES) was originally conceived as a means of utilizing electricity generated off peak to store energy for later release and generation during peak demand or on-peak periods. CAES is a technology that converts pneumatic energy by means of a compressor for later release through a gas turbine to generate electrical power. This arrangement differs from the conventional gas turbine in which the compression and expansion processes are physically connected with the compressed air passing directly into the gas turbine combustors. Moreover, by heating up the air released from storage prior to entering the turbine, the power output of the turbine can be increased significantly so that not only losses during the compression and pumping are compensated but also substantially more energy, in the order of 25 percent, is produced than stored. Since a hydroelectric pumped-storage plant has no potential means for compensating losses and for increasing the available energy, its energy ratio is always less than unity, that is, in the order of 75 percent. Thus, for a given available loading energy, a CAES plant can produce approximately 1.75 times more energy than a water storage plant. The first commercial CAES plant (290-MW capacity) was built at Huntorf in northwest Germany in the 1970s and has proven to be a useful adjunct to the electrical power system (Herrmann, 1975).

The application of a CAES plant to retiming the raw energy output of a tidal-electric plant is discussed by Haydock and Warnock (1982). A study to examine the economic viability of using CAES to retime the cyclic power from a proposed tidal-electric plant at site A8, at the mouth of Cumberland Basin in the Bay of Fundy (Fig. 16.8) was prepared for Energy, Mines and Resources Canada by the Canadian Atlantic Power Group (1988). The results of that study demonstrated the feasibility of retiming raw tidal energy output using a compressed-air energy storage system based on readily available equipment and methodology. The preferred system would be to use separate, electrically driven compressors and gas-turbine generators with variable pressure or uncompensated underground air storage. However, it was concluded that a tidal-electric plant/CAES system would not be attractive because of the fuel prices prevailing at the time of that study; the only gas turbine fuel available in the region at the time was distillate. By the end of the twentieth century, natural gas was available at a price about one-third that of distillate.

11.3.4 Auxiliary Capacity

Combustion turbines, basically jet engines connected to generators, could also be used to firm up a tidal plant's output. These turbines run on natural gas or distillate oil. Their advantage is that they can be started within a few minutes. However, because of normally high natural gas costs, they have been used primarily for peaking and standby reserve service. Following the first oil crisis of 1973–1974, fossil fuel prices increased substantially but, with the collapse of international oil prices in 1986, it again became economical for electric utilities to use oil and natural gas for electricity generation. Depending upon the availability and price of natural gas in the region of a tidal plant, combustion turbines could be used to firm up the tidal ouptut by generating when that output is insufficient or zero. For this alternative, the principal design variants would be the optimum fraction of tidal energy to be firmed up, the load factor of the firm product, and the duration of firmness required during the solar day.

11.4 SYSTEM CONSIDERATIONS

As with the conventional generating sources, investigation of the feasibility of tidal energy development requires studies of alternative generation expansion programs, including the incorporation of the new facilities into the power network (Karas, 1977). Alternative generation plans to meet the probable load reguirements up to 20 to 25 years into the future should be projected *with* and *without* a tidal power facility, in view of the long lead times for the construction of new plants, particularly hydro and nuclear (refer Table 11.1). A comprehensive approach is needed to develop alternative plans so as to have realistic mixtures of types and sizes of generation facilities. The tidal power scheme would then be incorporated into the expansion plan to determine when and how its output could best be utilized. A comparison of alternatives provides an assesssment of the relative worth of tidal development to the system. It should be emphasized that such comparisons must be made on the basis of maintaining the same degree of system reliability and security for each alternative.

This approach is not unique to utility practice in developing alternative plans for a system. However, until about the mid-1970s, investigations of tidal energy exploitation were based on the approach of developing an "equivalence" between the output of a tidal-electric plant and other conventional energy and firm power sources. Hence, the surfeit of proposals relating to two- and even three-basin schemes, as noted in Chapter 5. Such approaches necessarily placed the tidal power resource at a significant cost disadvantage.

TABLE 11.1 Approximate Lead Times to Commissioning of Large Nuclear and Fossil-Fueled Thermal Electric Plants

Steps	Nuclear 4 × 850 MW	Fossil 4 × 750 MW	Steam 4 × 200 MW
A. Site identification and approval to purchase process	2–3	2–3	2–3
B. Site investigation, preliminary engineering, and project approval process	3	3	3
C. Site preparation	1–3	1–3	1–3
D. Detailed design and construction up to first unit in service	5.5	4.5	3.5
Total time in years	11.5–14.5	10.5–13.5	9.5–12.5

Source: Mayers, (1980).

11.4.1 System Simulation

To carry out the systems approach, it is necessary to simulate, in detail, the utility system or systems so as to yield system costs of production over time, *with* and *without* tidal power. Therefore, it would be prudent and effective to develop a model to simulate the operation of the existing system and to select a generation schedule that supplies power when required, according to the expected load demands, at minimum cost. Data required are load forecasts and technical data that describe the operating characteristics and the costs of operation of all generating plants. The system would then be expanded to yield the least costs of energy and capacity in meeting the demand projected for future years. Committed or definitively planned expansion programs would be introduced first, followed by further expansion programs on the basis of the load forecast, technical and cost data of new conventional generation plants, and the following planning criteria:

1. *Planning Period* The delineation of this period should recognize the committed expansion of the system, the "lead time" for the various generation sources, the long-term economic effects of various expansion programs, and the fact that the effective life of a tidal-electric plant will be at least 60 years, similar to that of a river hydro plant.
2. *Restrictions on Generation Sources* Governmental policies may restrict or result in reductions in the use of some generation sources in the future.
3. *Reliability and Reserves* Generation planning requires the inclusion of an adequate generation reserve to allow for normal maintenance and unplanned breakdown of equipment. Required reserves have traditionally been expressed by various rules-of-thumb based on experience, such as a percentage of the minimum capacity required to meet an expected peak demand at a particular time; the percentage would vary according to system characteristics. For example, for smaller systems a frequently adopted rule was that the reserves must be equal to the capacity of the two largest units in that system. A more sophisticated approach is now in general use, that is, an index that reflects the probability that an expected peak load at the generation plant cannot be met. The interpretation of this *loss of load probabilty* (LOLP) index is rather complex with different indices being used by different utilities. Many North American utilities use an LOLP index of one day in 10 years. This means that sufficient reserves should be available to guarantee that a lack of capacity to meet the expected peak load would occur during not more than one day in 10 years on the average. It should be noted that the index qualifies neither the duration nor the severity of the shortfall. In smaller systems, this may mean fairly high reserve requirements of up to 25 percent—and even higher—particularly if very large nuclear units were introduced to the system since such units have high initial forced-outage rates relative to system capacity. Obviously, this would be a very costly reserve margin for smaller systems and an LOLP of one day in one year is often used provided that a reduction

in load during certain emergency conditions can be achieved so that the LOLP would be equivalent to one day in 10 years.

4. *Load Following Capability* The load-following capability of a system and the cycling of thermal units was discussed in Section 11.2.2.

The expansion of the system follows a least-cost path subject to technical constraints reflected in the foregoing criteria. The simulation of the system operation should be able to provide the least-cost generation schedule for any particular year. The total costs of the system over the planning period, consisting of the least cost of investment in plants and other related facilities and the cost of operation, would be obtained by calculating the *present worth* of all relevant costs over the planning period with appropriate discounting methods. When applying the simulation to the system with a tidal power plant, only the transmission costs directly associated with a particular tidal scheme need be included in the comparisons; transmission costs, which would be the same for both the *with* and *without* cases, would not be included. The costs of constructing and operating the tidal plant are not included in the simulations but all other relevant costs are. A flowchart for benefit–cost calculations is shown in Fig. 12.1, which indicates the point at which those costs are brought into the analysis.

This same approach would be used to examine variations of a particular tidal power scheme, for example, with and without pumping, double-effect, and so on, or various export alternatives.

11.4.2 Market Analysis

Ideally, the allocation of tidal energy between a primary and a secondary market should be determined by optimization. However, the problem is complex and a sequential approach to market analysis, illustrated by Fig. 11.6, would allow system planning of the market areas to be done independently (Lee and Deschamps, 1977). The primary market is given the priority in the direct absorption of tidal power. The displacement of nuclear generation is dictated by the technical capability of nuclear units to regulate their output. By varying the assumed must-run nuclear generation, different allocations of tidal energy to the markets can be simulated. Storage devices are obviously one way of increasing the utilization of tidal energy in a market area. By varying the transmission capacity between the markets and the amount of storage in the primary market area, an optimal combination of storage and transmission capacities can be determined.

11.4.3 Simulation Models

In an engineering economic analysis of tidal power, benefits are of two types: fuel cost savings and generating capacity savings. Under a state of high fuel prices, fuel cost savings would be the more important so that the best market

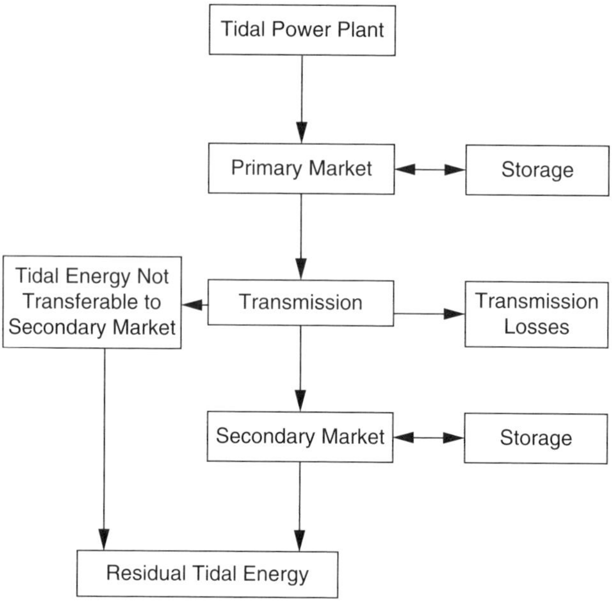

Figure 11.6 Sequential approach to market analysis.

for tidal power would be one that has a high proportion of fossil-fired generation and that is located near a tidal site to minimize cost of transmission. If the capacity of the tidal-electric plant is large, its marginal value to the primary market decreases because tidal energy begins to displace relatively inexpensive nuclear energy or the market cannot absorb the total tidal plant's output because of technical constraints. If the latter constraints govern, the introduction of secondary or even tertiary markets, depending upon the cost of transmission between the primary and such markets, should be considered since the difference in the respective values of tidal energy could more than offset the transmission cost and so increase the total savings.

The models to carry out the procedures described in the foregoing should provide for two types of functions: *planning* and *simulation*. The planning functions are generation mix optimization and determination of reserve requirements, while the simulation functions are generation dispatch of nonthermal resources and thermal production costing. For example, the relationship between the computer models that were developed to carry out the planning and simulation functions for the Bay of Fundy (Canada) reassessment study is shown by Fig. 11.7. In essence, the models are used to develop a least-cost expansion plan without tidal power. Then, by introducing a tidal-electric plant, with or without supplementary storage, into the expansion plan and allowing for an appropriate lead time, a different expansion plan is derived that is compared automatically with the reference plan to give the benefits attributable to tidal power. This approach proved very effective and productive and, although

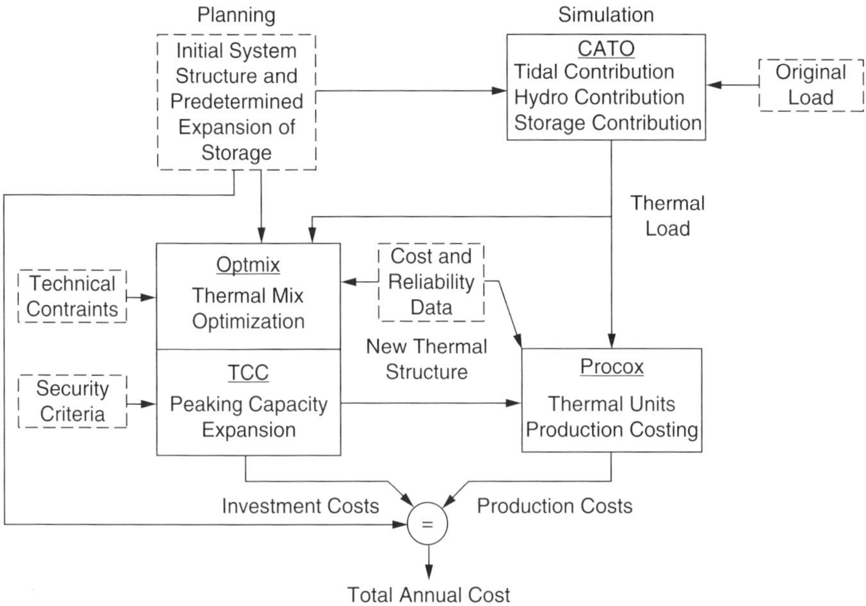

Figure 11.7 Functional relationship of computer models to analyze technical and economic viability of alternative generation expansion programs.

such models may now have reached the third or fourth generation of development, their logic is still valid when the generation is owned and operated by a franchised utility (Lee and Deschamps, 1977) and is described briefly to illustrate the various aspects of the planning and simulation functions.

Referring to Fig. 11.7, the model simulating the Chronological Absorption of Tidal Output (CATO) determined the contribution of nonthermal generation sources to supply load and computed the residual load called thermal load. This provided input to both the OPTMIX (Optimum Mix) model for determining the optimal thermal mix and the TCC (Total Capacity Correction) model for determining the reserve requirements. As a result, a new thermal generation structure was determined. CATO also performed an iterative process until an optimum structure was determined. The Production Cost (PROCOX) model was then used to compute the thermal production cost.

Due to uncertainty of important data and the inherent computational burden, dynamic optimization of the expansion plan may not be practical. For the Fundy studies, a static optimization was performed using a "snapshot" year approach with about 5 years between each snapshot of the expansion plan. Interpolation was used for intermediate years by a FILLIN model, that is, a model designed to take the investment decisions made by OPTMIX and TCC for the current snapshot year and to interpolate them over the intermediate years. The production costs for the intermediate years were then computed by PROCOX. Upon exit from FILLIN, the system structure from the previous snapshot year to the current snapshot year was linked and the associated fixed charges and annual production costs for the intermediate years determined. By properly spacing the snapshot years and interpolating the results for the intermediate years, it was possible to develop a near-optimal expansion plan.

The central model, CATO, determined on an hourly dispatch basis over a one-year simulation period, that is, for each snapshot year, the tidal absorption capability within the primary market and subsequently within the secondary market. The objective of this model was to determine in each snapshot year the generation schedule of all the nonthermal resources in order to identify the residual load that was to be supplied by the thermal system.

The model functloned by first dispatching all nonthermal generating sources such as tidal output as well as conventional hydro and pumped-storage devices to meet the hourly system loads. In carrying out this task, such constraints as must-run generation (i.e., uncontrollable hydro, nuclear, and fossil units, as discussed previously) were taken into account. All available tidal generation was first utilized in the primary market to maximum advantage. Any surplus generation was then allocated to the other markets. The residual hourly loads remaining after dispatch of all nonthermal generation were then arranged in ascending order to form load duration curves (LDCs). The thermal load–duration curve without tidal output and termed the *original LDC* as well as the *LDC modified by tidal input* was then used as input into the two models—OPTMIX and PROCOX.

The model OPTMIX selected, for the planning process, the optimum mix of nuclear and cycling thermal generation based on capital and operating cost screening curves while the optimum sizing of pumped-storage capacity was obtained in a deterministic manner through iterations with the OPTMIX and CATO programs. As indicated previously, the subprogram TCC of OPTMIX was used to determine the necessary additions of peaking units, taking into account the forced outage rates of all thermal generation, to meet the required level of system reliability measured in LOLP. The program, therefore, allowed for consideration of the capacity contribution of a single-basin tidal plant to the system load-meeting capability. With the reliability criteria properly satisfied by the generation mixes with and without tidal power, the capacity credit of tidal power was correctly factored into the economics.

The resulting thermal generating capacity from OPTMIX and the associated operating costs as well as the original and modified LDCs were then used in the PROCOX model to calculate the total annual production cost of thermal generation for the comparison of alternatives *with* and *without* tidal generation.

Representative small, medium, and large tidal plants should be fed into the system modeling analysis process to determine when and how the output could best be utilized in the electrical utility system and contiguous market systems as well as the resulting modifications on the generatlon expansion programs. As noted previously, tidal energy absorbed within the interconnected systems has the effect of displacing or reducing a like amount of conventional thermal

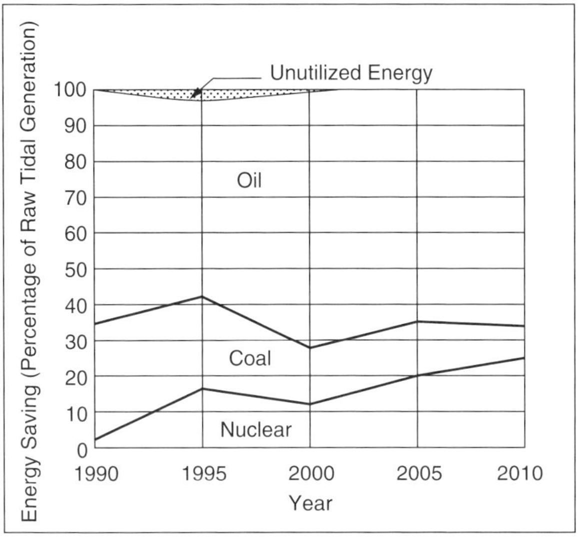

Figure 11.8 Example of energy savings by fuel source to electrical utility system with a single-basin, single-effect tidal plant.

energy production within those systems. A significant portion of the value of tidal energy can thus be measured in terms of the amount of thermal energy displaced and its type of fuel. This result can be illustrated in the manner shown by Fig. 11.8 in which the values of energy savings were derived by optimizing the expansion plan at 5-year intervals with a linear interpolation between the snapshot optimization years.

CHAPTER 12

ECONOMIC EVALUATION

12.1 INTRODUCTION

The determination of the economic feasibility of tidal power, that is, its competitiveness over the long term with conventional sources of energy used by utility systems, has been shown in the previous chapter to involve the interaction of many factors under future conditions that cannot be predicted with certainty. The primary objective of an economic analysis is to identify the generation expansion programs that will provide a required service at the least cost. Thus, the economic evaluation of tidal power is based on the cost comparison between an expansion plan *without tidal power* and one *with tidal power*. For the latter expansion plan, the cost of the tidal-electric plant is not included so that the cost difference of the two plans is the total benefit due to the introduction of the tidal plant. This value would then be compared with the cost of construction of the tidal-electric plant to complete the analysis.

The direct costs of the plant are the costs of construction, maintenance, and operation, and its direct benefits are the contributions that its output makes to the utility system. The quantification of the benefits of a tidal-electric plant to an electric utility system is complex and is determined by modeling the utility system as described in Chapter 11. This approach takes account of possible changes in system operating costs and the pattern of investment resulting from the introduction of the tidal-electric plant since it also illustrates the least-cost mode of meeting the electrical demand.

Elements of Tidal-Electric Engineering. By Robert H. Clark

Of course, there will also be indirect costs and benefits due to the impact of the development on the environment and on the activities of society in and around the estuary. The direct costs are relatively easy to determine, but the indirect effects are more difficult to quantify.

12.1.1 Assumptions

As a basis for an economic evaluation, it is necessary to adopt several economic parameters and assumptions, such as: (1) the selection of the reference date to which all costs are to be referred; (2) an internally consistent method of accounting for the time value of money, general inflation, and fuel price escalation; and (3) a method of overall comparison of costs and benefits that, typically, will vary from year to year over the project life. The following basis is recommended:

1. It is usually convenient to select the project in-service date as the reference date for costs. Costs incurred at other times are discounted, or compounded, to the reference date using an appropriate discount factor.
2. In an economic evaluation, it is convenient to use a *real* interest rate as the discount factor. The real interest rate accounts accurately for the difference between the actual interest rate and the general inflation rate, with the result that general inflation can be ignored and calculations can be made in constant dollars. It has the further advantage that the results obtained are not affected by changes in the actual interest rate or the inflation rate, so long as such changes do not alter the real interest rate. However, any divergence between the general inflation rate and the rate of increase in fuel prices must be accounted for by applying a differential escalation factor to the latter.
3. A convenient basis for comparison of costs, or benefits, that change over time is *levelized* cost. This is the constant cost, invariant from year to year, which would have the same present value as a given series of varying annual costs. Annual costs and benefits of a project can be expected to change from year to year, perhaps in complex ways. Cost–benefit comparisons for each year would, under such circumstances, be difficult to interpret, whereas the comparison of levelized costs yields a single figure to provide the comparison.

12.1.2 Definitions

Real Interest Rate This is the approximate difference between the actual interest rate (taking into account the borrower's credit rating, the risk involved, and other related factors in the venture) and the current inflation rate so as to exclude the general inflation effects on alternative fuel prices.

Levelized Cost The levelized cost of energy production over a specified period of time is determined utilizing price discounting procedures. As used herein, it is the present value of the production cost of the energy that is displaced by tidal generation energy divided by the amount of energy displaced, discounted at the same rate. In algebraic terms it may be defined by the equation:

$$\text{Levelized cost} = \sum_{i=1}^{i=n} \frac{C_i}{b+dg} \div \sum_{i=1}^{i=n} \frac{E_i}{(1+d)^i} \tag{12.1}$$

where E_i = energy displaced in year i
C_i = production cost of energy displaced in year i
d = discount rate

A tidal-electric development is considered to be *economic* if the levelized benefit to levelized cost ratio exceeds 1. The *breakeven* point of the development would occur at the point in time when the total present worth of benefits equals the total present worth of costs, that is, the period for the benefits derived from the tidal plant to cover all the associated costs of the development. In this connection it should be noted that the economic life of a tidal-electric plant is more than three times that of a nuclear power plant or a fossil-fired thermal power plant.

Present Value Present value represents the value, in a base year, of a dollar value in any other year or a series of dollars valued over a number of years, after adjustment for the effects of price inflation. *Present worth* and *discounted value* are often used as synonyms for present value. The present value V of an amount A in year n at discount rate d is given by the formula,

$$V_n = A(1+d)^{-n}$$

The present value of a constant annual cost or benefit, A, per period for a term of n periods at a rate of interest d is given by the formula

$$a_n = \frac{A[1-(1+d)^{-n}]}{d}$$

It should be noted that comparisons of the present worth dollar value of the benefits and costs, or the equivalent levelized mills/kWh amounts, with the current cost of alternative generation, are not valid. Present worth values are only indicators of the relative magnitude of benefits and costs associated with each tidal development, but the ratio of these quantities is an important indicator of economic viability.

12.2 ECONOMIC COST PARAMETERS

12.2.1 Cost of Capital

The economic cost of capital is expressed in the form of an inflation-free discount rate. There is considerable literature discussing alternative concepts of the cost of capital that should be reflected in the choice of the discount rate to be used in evaluating projects in particular situations. One concept relates to the real cost of capital associated with the basic concept of social time preference valued for a particular time. The other concept relates to the productivity of capital in alternative investment opportunities. These two concepts should ideally be reflected by one discount rate, but this is rarely the case in practice due to the characteristics of the capital market and its imperfections. Various methods have been suggested to bridge the difference (Eckstein, 1961).

The problem, from the point of view of this text, is one of choosing between alternative methods of generating power, all with a long project life, and not one of choosing between investment in the power sector as compared to investment in other sectors of the economy. This would suggest that the concept of the "real" cost of capital, rather than the productivity of capital in alternative investment opportunities, is the more appropriate concept.

12.2.2 Cost of Labor

Major changes in the cost of labor could significantly affect the cost of construction of all power projects. The future cost of labor will result from real wage increases (adjusted for inflationary increases), changes in fringe benefits, and of labor productivity. A very large project could result in an upward pressure on labor costs in the region.

12.2.3 Cost of Fuels

Beginning in the early 1970s, the price of oil was manipulated by the Organization of Petroleum Exporting Countries (OPEC) cartel through restrictions on production. Subsequently, as a result of huge investments by non-OPEC countries, supplies were developed elsewhere in the world so that, currently, about two-thirds of the world's oil supplies comes from non-OPEC sources. Yet, it is interesting to note that a large proportion of the additions to reserves has been in the OPEC countries, which will undoubtedly pose problems for future prices.

Nevertheless, taking into account the market forces, the economic growth in the industrialized countries and in the rest of the world, the long-term effect of price on consumption, development of alternative sources of energy, and the committment of funds by industrialized countries to the development of new, relatively high-cost sources of hydrocarbons, the best estimate of likely future oil prices must be made to reflect the market forces, and the like. The estimate must also reflect the relative change in oil price with respect to general inflation in order to provide a base for a proper comparison of a power system with and without tidal.

The cost of coal is more likely to be determined by the cost of production, including transportation to consumers, than is the case for oil. The estimate of coal costs in the future should, therefore, be less uncertain than those for oil.

There is still considerable uncertainty regarding the recoverable reserves of uranium at various price levels.

12.3 ECONOMIC ANALYSIS

An economic evaluation of tidal power is based on the cost comparison between an electrical generation expansion plan without tidal power and one with tidal power. A flowchart for benefit–cost calculations is shown in Fig. 12.1. It will be noted that the cost of a tidal-electric plant is not included in the costs of an expansion plan with tidal. Therefore, the cost difference of two such expansion plans would represent the total benefit due to the introduction of the tidal plant. This total benefit could then be compared to the construction cost to complete the analysis.

The relevant costs for evaluating an expansion plan would be as follows:

1. Levelized fixed charge rates would be used to compute the annual carrying charges for the new investments on thermal generating capacity.
2. Fixed operating and maintenance (O&M) charges would be computed each year for the new thermal units only.
3. Annual production costs would include variable O&M charges for all thermal units (old and new).
4. Annual cost would be the carrying charge plus fixed O&M plus production cost.
5. Present worth would be computed for each year and cumulated.
6. When the last "snapshot" year has been reached, the load growth would be assumed to stop and the annual cost of the last year would be repeated for the additional years, termed the evaluation period (assumed in the case of the Fundy studies as 35 years) and their present worths would be added.

If a secondary market were involved, the transmission capacity would be optimized by simulating different voltage levels and including the transmission costs in the economic analysis. The total benefit due to tidal power will be given by Bx, where x is the transmission capacity and C_t is the transmission cost:

$$Bx = B_1 + B_2(x) - C_t(x) \tag{12.2}$$

Subscripts 1 and 2 denote primary and secondary markets.

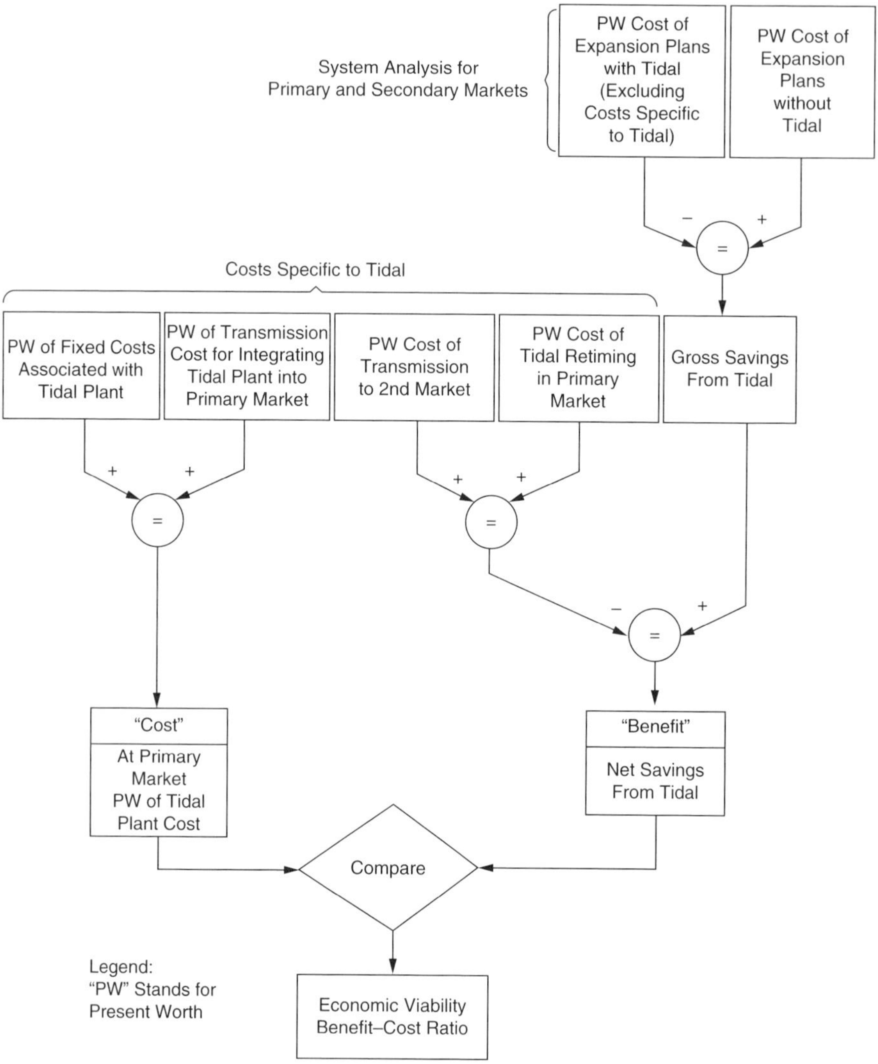

Figure 12.1 Flowchart for benefit–cost calculations.

The maximum value of B can be found by solving the equation for several values of x. The equation assumes that the saving in the secondary market is completely credited toward the construction cost of the tidal plant. If additional storage were considered for retiming tidal power for the primary market, the optimization of both the transmission capacity x and the storage capacity y would be performed on the following function:

$$B(x, y) = B_1(y) - C_s(y) + B_2(x, y) - C_t(x) \tag{12.3}$$

where C_s is the storage cost and other symbols are the same as those of Eq. (12.2). The problem is now two dimensional, but it can be solved by repeated simulations for discrete values of x and y.

12.4 SENSITIVITY ANALYSIS

A base-case economic analysis will likely be developed on what is considered to be the most likely scenario, that is, the delineation of a hypothetical but feasible future, using average assumptions regarding the values of parameters that may significantly affect the economic feasibility of tidal power. Because of the unpredictability of precise future values of parameters involved in a scenario, it is recommended that a range of values for each such parameter be included in the economic analysis. In this way, the effect of an individual parameter on the economic feasibility can be assessed. Such a sensitivity analysis provides as complete a picture as possible of the future consequences of including, or of not including, tidal power in the generation expansion program of a power system, which includes both the generation and transmission function.

Some of the parameters of an economic analysis that could influence the benefit–cost determination are:

1. Extent of nuclear development or penetration in a utility system expansion program
2. Load forecast changes
3. Marketing strategies for tidal energy in a secondary market
4. Fuel costs
5. Discount rates
6. Load factor
7. Costs of storage and of transmission
8. Cost of construction of the tidal plant
9. Growth of demand for energy

Of these, there are at least three parameters with uncertain and controversial future values that will likely have a significant influence on benefit and cost determinations. These are *discount rates, extent of nuclear-power penetration,* and *fuel costs.*

12.4.1 Discount Rate

The discount rate probably invites the most controversy of any parameter used in the economic analysis. As noted previously, the more appropriate concept to use in analyzing alternative expansion plans for an electrical energy system is the concept of a "real" interest rate. The discount rate exerts great influence on the benefit–cost ratio of a tidal power development because the cost of output consists almost entirely of the annual cost of capital while the benefits, derived largely from the displacement of thermal energy, are relatively unaffected by the discount rate. In other words a tidal-electric development is very capital intensive with capital amortization accounting for about 90 percent of the unit cost of tidal energy. Accordingly, the cost of construction is one of the major factors influencing the profitability of the project.

The choice of discount rate should be made within the context of the practice of the electric utility that will "own" the tidal-electric plant. A high discount rate would argue for the less capital-intensive fossil-fired plants and against capital-intensive nuclear and tidal-electric plants, a result that would not assist in the reduction of greenhouse gas emissions. However, it may be appropriate for a senior government or an international banking institution to use some other discount rate when considering the question of financial assistance to the project. In that connection, however, it should be pointed out that if the concept of the social opportunity cost of capital is applied, then the following should be kept in mind:

1. Benefits should also be calculated on a social basis and should include tax recoveries on incremental economic activity and employment created by choice of the more capital intensive alternative, multiplier effects, reduction in government welfare costs, resource conservation benefits, and the like.
2. Choice of social opportunity cost of capital should reflect the particular industry involved.
3. Social opportunity cost of capital varies from time to time.

12.4.2 Nuclear Penetration

Tidal power in its most economic role would displace fossil-fueled energy production and would likely have little effect on existing or the need for new nuclear generation. Therefore, the greater the amount of nuclear generation in the system, the smaller the scope for absorption of tidal energy. One of the basic objectives in a feasibility study is to determine an optimum generation expansion program without tidal power to minimize total system costs, but consistent with established technical constraints associated with various types of generation. Thus, the alternative generation programs would be derived to meet a prescribed reliability criterion. In the simulation study when nuclear generation is added to this system and energy surpluses are available, a corresponding amount of pumped storage or other types of storage devices could be introduced to the expansion program to utilize these surpluses to the best advantage. The storage devices would thus form part of the total generation mix. Nuclear plants now in service are designed for base load operation. While the load following characteristics of the turbines do not pose any problems, the response of the reactor is much less flexible so that a sudden reduction of reactor power to less than 50 to 70% of maximum continuous rating will result, after about 30 min, in reactor "poison out." Once the reactor is in that state, it takes about 26 h for the waste products (i.e., xenon) to decay before the reactor can be restarted. In terms of load following characteristics, the output should not be reduced to 75% of maximum continuous rating on a daily basis. The differential form of stress on the fuel bundles and on the reactor itself are such that an increase in failure rate could result from this type of operation over a long-term period. Once such a failure occurred, the plant could require shutdown and incur costly repairs. Nuclear has been, and is still, held by many experts to provide the key to energy supplies (Chamberlain, 1989), but the fulfillment of this expectation may depend upon the trends of environmentalist intervention and the price of nonnuclear energy. In North America, combined cycle generation and, where thermal demand exists, co-generation are expected to dominate for the forseeable future.

12.4.3 Fuel Costs

The value of tidal energy is very sensitive to fuel costs since a significant portion of the benefits would derive from the displacement of fossil-fueled thermal energy. The marked changes in price of a barrel of crude oil, commencing in 1973, have demonstrated the volatility of the price of this commodity and the tenuous nature of the projection of future long-term oil prices. One can only rely on the advice of the experts in this field and on specialized studies as a guide in the selection of the "most likely" fuel cost scenario adequately supported by alternative scenarios to provide guidance on the effects of the inherent uncertainties.

12.5 RISK ASSESSMENT

The profitability of any energy project turns fundamentally on the expected prices for future energy. It should be kept in mind that demand is quite inelastic, which was illustrated when OPEC quadrupled oil prices in October 1973 but world demand fell only by 5 percent. On the other hand, a small oil surplus can cause a major price fall. Future additions to nuclear capacity, which has lower variable cost than other thermal sources, may well depend upon the weight of environmental opinion and the price of nonnuclear energy. The price of energy is a vital component in the economics of a tidal power project so that known forecasts of future oil and energy prices are basic to an investigation and require continuous monitoring. A general view of the electrical energy situation in 1963 is provided by the following statement (Whipple, 1963):

> Hydro-electric power development is also levelling off since most of the larger favourable reservoirs sites have already been developed, or so utilized for other purposes as to make hydro-electric development impracticable. Moreover, the day when atomic energy will become cheaper than steam may be less than 20 years away. Already we should be discounting the projected value of power to be produced from hydro-electric power projects.

The failure of the Three-Mile Island nuclear power station in 1979 and the nuclear reactor disaster at Chernobyl in 1986 were major setbacks to the expansion of nuclear power generation. During the last two decades of the twentieth century almost 100 nuclear stations globally have been shut down or decommissioned. The foregoing illustrates uncertainties in forecasting the energy picture.

Every project has to be planned and carried out within the existing contexts of certain physical, commercial, social, political, and economic circumstances. Within each of these, there is an area of uncertainty surrounding major projects from the conceptual plan to final completion. A substantial aspect of the management of a project involves attempts to impose control of these circumstances. Close examination and careful consideration will show that these areas are made up of specific risks, viz. *events* that can occur to the detriment of the project.

Obviously, an event that would be unlikely to occur but that could cause losses of many millions would require more attention than one that would be quite likely to occur but could only cause losses of a few thousands. Risk may be defined as the likelihood of the occurrence, that is, the *degree of probability,* of a definite event or combination of events that could occur to the detriment of the project. Risk is not the generalized uncertainty with which it is often confused. It is not only advisable but essential in major projects to be absolutely precise as to the nature and extent of all significant risks involved and plan for each of them. This is the only way in which the project can be protected and the ultimate residual risk quantified. Linked to the risk would be the *amount at stake*, that is, the total loss or losses that could result. In large projects, the risk minimization technique must identify and quantify the individual risks and recommend the most economic response to each of them, taking into account the many specialized disciplines involved, (Fraser, 1979). Risk identification (Fig. 12.2) is a continuous process from the prefeasibility

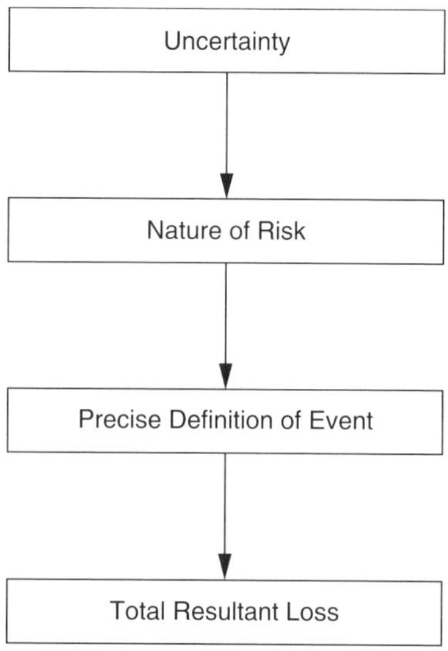

Figure 12.2 Risk identification and quantification. (From Fraser, 1979.)

stage to project completion, together with the appropriate assessment and analysis of the risks (Fig. 12.3). If the residual risk is unacceptably large, then the project is not viable; if it is acceptable, then the application of detailed analysis and planned controls and security gives the necessary confidence to all parties concerned.

In feasibility studies it is necessary to assess, as accurately as possible, all of the risks and uncertainties that may arise in undertaking the development,

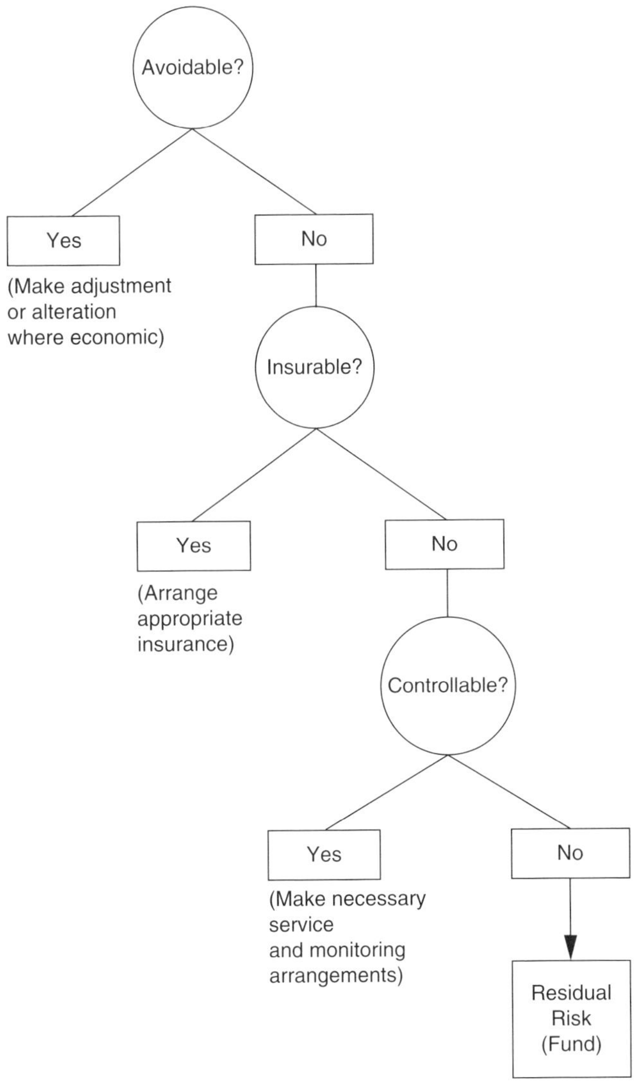

Figure 12.3 Risk analysis. (From Fraser, 1979.)

bearing in mind the lead time necessary for commissioning such a project. The sensitivity analysis outlined in the previous section may also be applied to some of the following risk factors, although in some cases, with much less precision.

Design of the project defines its form and substance. The initial plans, as well as the approaches to meeting them, involve risk. For example: Is the design based on proven technology? Is the scale of application appreciably increased? Is the project scope adequately defined for a reliable capital cost estimate? Engineering management has the responsibility to ensure that the project is built in optimum fashion to meet design and financial requirments. Many other factors play important roles, such as construction, manufacture, quality assurance, commercial risk assessment control, insurance, risk analysis and contingency determination, contracting policies and project management of the total enterprise. Close integration of the engineers' and management's roles has proved to be a most effective method of controlling risks and of achieving the demanding goals set for large projects. To a substantial degree, potential design risk can be kept in line by effective engineering management (Warnock, 1979). The important role of project management is dealt with in detail in Chapter 4.

The main areas of risk in the economics of a tidal-electric development are likely to be delays in the barrage construction program and delays in turbine manufacture, installation, and commissioning (Severn Barrage Committee, 1981a). The value of the energy produced for the electrical system in which it would be utilized, through benefits, would likely be even more uncertain than the costs because of the uncertain future and the need to look a long way ahead. High-caliber management is required to manage any and all possible risks to ensure that their effects are minimized. One shouldn't be misled by the term *risk management*. It does not imply control of events; rather, it is a response to such events in advance or as they occur. The important role of project management and the type of management team required for tidal power feasibility studies is dealt with in Chapter 4.

On the one hand, it may be argued that conventional power sources are relatively devoid of technical risk, and construction costs, being reasonably controlled, can accordingly be predetermined with reasonable accuracy. On the other hand, cost estimation for new technologies is likely to be a difficult and uncertain task (Merrow et al., 1979). For example, in the case of a Bay of Fundy project, where the physical environment may be termed hostile and construction methods cannot lead to a narrow range of forecast of capital costs, problems will be posed that differ in size and scope from ordinary "big" projects and demand a management of the highest calibre to ensure that the risks are minimized and the project optimized (Sykes, 1979).

CHAPTER 13

SOCIAL AND REGIONAL IMPACTS

An evaluation of socioeconomic impacts is an essential part of an overall appraisal of a tidal power development. The potentially significant issues should, at least, be identified by a prefeasibility study, and those aspects requiring detailed analysis should be highlighted for subsequent study.

The social and regional impacts of a development would be similar to those of other capital-intensive projects, particularly to those of a large hydroelectric project. The investment period, characterized by very high levels of activity and expenditure, would be followed by a long operational period during which these levels would become quite low. Capital charges would amount to 90 percent of annual costs, or perhaps more, with operation, maintenance, and replacement accounting for most of the balance. The costs of maintenance and replacement would be quite small compared to these capital charges; other costs of operating the facility would be almost negligible. The impact of the project may be expected to be very large during the period of construction and quite small during the long period of operation. Escalation in these costs number, however, be taken into consideration.

For a prefeasibility study it should only be necessary to limit the investigation of these impacts to the exploration and identification of potentially significant issues. Such issues would be the subject of detailed analysis during subsequent studies.

However, a tidal power development also presents certain aspects and options that are very different from more conventional power sources and which could yield distinctly different social and economic results. The

Elements of Tidal-Electric Engineering. By Robert H. Clark

following are examples of the uniqueness of a tidal power development from the socioeconomic point of view:

1. Storage and generation take place in a marine environment so that the development would require only a small amount of land permanently, such as for access roads, connecting public roads, electrical substations and transmission lines, maintenance yards, control buildings, and the like; consequently, few, if any, relocations of people will be required and very little reallocation of land and water.
2. A likely construction option will be the floating in of prefabricated caissons and sinking them on location as components of the barrage; the adoption of this method would permit a significant amount of the work to be done remote from the site.
3. The available energy at a tidal site may require the installation of large numbers of turbogenerators, a number that may be large enough to justify establishment of a local industry for their manufacture and overhaul; for example, 216 turbogenerators are proposed for the Severn project (Severn Tidal Power Group, 1989) and 420 for the Tugur (Russian) project (Bernshtein, 1992).
4. Since tidal energy is generated in surges lasting from four to six hours, with intervals of about eight to six hours of no output, energy intensive industries that could work on that rythm of power availability may find the region of the tidal-electric plant an attractive location.

Since a tidal-electric barrage would be constructed at some location along an estuary, it could have significant impact on ports and shipping in the estuary, whether because navigation locks would be required, or major changes would be manifested in the water level regime, particularly if the port were upstream, or because of changed sedimentation patterns. Any such impacts on the stream of economic activity in the region would, of course, require assessment with the possibility of introducing remedial measures to maintain the current activities or to enhance them. The barrage could be designed to serve also as a causeway across an estuary that could have significant impact on traffic and communities on both sides of the estuary.

Direct onsite employment estimates and the muliplier effects of this in the economy and social structure of the region and, indeed, on the nation, require quantification and evaluation. If the unemployment rate were high at the time construction starts, this would provide an additional benefit to the project; on the other hand, full employment in the region could result in higher labor costs.

The peak labor requirements would be of the same order of magnitude as for a comparable-sized hydroelectric site and would be only slightly higher than the average requirement over the construction period. By careful scheduling of the work, it should be possible to arrange for a relatively steady level of employment throughout the construction period. If it were desirable to reduce onsite employment to minimize impacts on the social and physical infrastructure of the surrounding area, the use of floated-in caissons would make this possible. The construction of such caissons could be carried out away from the site at a suitable location, even a relatively long distance from the site, if necessary (refer to Chapter 14). Supplementary labor requirements, in addition to the direct onsite requirements would be of two types. The first would consist of labor employed in the production of supplies, such as cement, concrete, lumber, aggregate, steel products, turbines, generators, and other electrical products. Some of these activities would not likely be located in the impact region nor even in the country. The second type of supplementary labor requirement would relate to the additional jobs to supply the demand for services by the labour employed onsite and in supply activities; the number of the latter type may be in the order of about one-half the direct labor force. If the construction of a tidal power project were located in an essentially rural area containing several small towns or villages, careful planning would be necessary to reduce the disruptive strain of a "boom–bust" nature in the immediately surrounding rural areas and to spread, to the extent possible, the demand for services and facilities over a wider area.

The materials, intermediate goods and equipment required for a tidal project may be grouped into three main categories: (1) raw materials such as aggregate, rock, cement, and lumber; (2) steel products, including reinforcement and fabricated gates and stoplogs; and (3) electromechanical equipment including the turbogenerators, transformers, and switchgear.

A survey of those potential manufacturers and suppliers within the region of the potential project, and subsequently within a widening range, would determine, at least on a preliminary basis, the most economic and reliable source for the foregoing categories of materials and equipment. Such a survey should also yield some intimation of the prospect of establishment of a new industry in the region (or in the country) to supply some of the manufactured products. Taking into account the foregoing considerations, a materials-origin supply study and an evaluation of the labor supply and its limitations should also be developed in a prefeasibility study.

A tidal power development has generally been considered in terms of relatively large installations and would, therefore, involve a substantial investment. This may or may not be significant in the national context, but it should be borne in mind that the construction period, including the preconstruction activities, would likely be on the order of a decade. Although such a project may not have an effect in the national context, it would undoubtedly have a substantial impact in the regional socioeconomic sense and lay a substantial claim on the economic resources of the region in which it is located. The temporary construction activities could provide opportunities to strengthen the local infrastructure and provide lasting benefits. For example, access roads would be required and/or existing roads upgraded.

The very substantial increases in the price of fossil fuels during the 1970s made it desirable, in those countries with adequate fossil fuel substitutes, to invest heavily in very capital-intensive nuclear power plants. Investment in a tidal-electric plant would imply a further substitution of capital investment for future expenditures on fossil fuels and would reinforce the general tendency to a capital-intensive generation system. The implications of an electric power system expansion, with and without tidal power, will require careful consideration in the context of the expected investment activity in the region.

The impact of a tidal power project on the balance of payments of a nation may be separated into the short-term impact experience during the construction period and a long-term impact during the operational life of the plant. The short-term impact would depend primarily on two factors: the relative proportion of investment funds that would have to be borrowed outside the country and the value of equipment and materials that would have to be imported directly or indirectly. Over the long term, foreign exchange would be conserved through the reduction of energy imports, although the repayment of foreign loans would represent an outflow of resources. Overall, if a tidal power development is found to have a positive benefit/cost ratio, it is likely to reduce foreign exchange payments.

There are undoubtedly other factors to be considered in evaluation of the feasibility of a tidal power development depending upon national energy policies and requirements. Some of them may be relevant factors in the decision-making process but are not likely to affect significantly the justification of such a development.

CHAPTER 14

ENVIRONMENTAL ASPECTS

The decision to build a tidal-electric generating facility must be made with the knowledge that changes will be made to the environment of the region surrounding the estuary and possibly also to areas seaward of the barrage. The basic questions are: How will the facility affect the environment? Beneficially or detrimentally? How can one tell? And what can be done about those effects that are detrimental or will increase the stresses on the indigeneous ecosystems?

The evaluation extends beyond mere economic analysis of production of electrical energy to value judgments on social contributions relative to social costs. The environmentally conscious public will also be disturbed and confused. With much emotion and little knowledge or reason, recently arrived ecological and environmental missionaries may mislead the people (Clark, 1997). There is a danger for an emotional approach to ecology to ignore man and his needs and aspirations as a legitimate part of the ecological system. A rational approach implies that first there be a realistic concept of conservation. Interest in the environment should not be for its own sake but for the sake of the people who must live in it; irrationally, some hold that conservation is preservation, forbidding use. Nonuse is total waste and is as much a loss as overuse. The rational concept is balanced multiple use with the objective of fullest use over the years within the total ecological system, without waste or abuse. With imagination and innovation, large, multipurpose, tidal-electric generation facilities, including pumped storage developments, have great

Elements of Tidal-Electric Engineering. By Robert H. Clark

potential as a dual power–water system to improve the environment and enhance the quality of life (Karadi et al., 1971).

The construction of a barrage that alters the regime of tidal movements will have an immediate and direct effect upon the environment of the intertidal zone and upon the animals and plants that live there. Regrettably, the existing database of an estuary with tidal power potential is usually meager and considerable subjectivity has to be used to identify the envionmental issues. An excellent overview of the environmental effects of tidal energy is presented by Muirhead (1992). As pointed out by Gordon and Longhurst (1979), impact assessment studies usually have to be concluded simultaneously with, rather than subsequent to, the required basic environmental studies and, as the latter proceed, some of the presently recognized issues may well change, requiring some redirection of the impact assessment studies.

Fortunately, more attention by research institutes has recently been focused on developing basic environmental study programs for estuaries, particularly those that appear to offer some promise for energy exploitation. Several workshops, seminars, and symposia have also been organized within the past two decades to exchange information and advance the debate on estuary management (Daborn, 1977; Severn et al., 1979). Shaw (1980) has edited about 18 articles on environmental factors with reference to the Severn Estuary in the United Kingdom (see also Severn Barrage Committee, 1981a, Vol. 2, Chapter 14, and Severn Tidal Power Group, 1989, General Report, Section 3.). The workshop held at the Université de Moncton in November 1982 reviewed the then current knowledge of possible environmental impacts of large-scale tidal power developments in the Bay of Fundy, and the proceedings provide an excellent basis for planning additional studies of the environmental components (Gordon and Dadswell, 1984). The reader is also referred to the publications listed in the Bibliography section.

A set of draft guidelines for the preparation of an environmental impact statement was prepared by the Bay of Fundy Tidal Power Environmental Assessment Panel (1978) to reflect the requirements of the environmental impact policies and procedures of the governments of Canada, New Brunswick, and Nova Scotia relating to a Fundy tidal power project. These guidelines were intended to define for the proponent, the scope, orientation, and general format of an environmental impact statement (EIS) and at the same time provide guidance on the particular concerns or issues that should be addressed. In the introduction to the draft guidelines, the panel stated:

> Given the possible wide ranging consequences of a project of this magnitude, the guidelines stipulate that the E.I.S. must forecast effects or changes within the social/community sphere as well as the natural or non-human environment. Within this expanded concept of the environment, the guidelines are invited to specifically orientate the E.I.S. toward the identification of likely changes in existing patterns of resource uses, functions and potentials. In order to identify and predict the direction and magnitude of these changes, an understanding of

the natural and human resources of the region influenced by the project will be obtained and presented.

Every major technological advance through history has carried with it new obligations to guard against misuse and that is vitally true of tidal-electric developments. It is obviously important to be able to predict the biological response to such changes and to assess their ecological significance because, if they are sufficiently important, the predicted ecological consequences will have to be taken into account in the design, construction, and operation of any barrage system. Undoubtedly, as for other types of power stations, some schemes will be more environmentally acceptable than others; moreover, this will not be appparent without the comprehensive studies required for any large or giant project exploiting a renewable resource. The purpose of this chapter is only to identify the more important of the environmental issues and not to develop a consolidated statement on any one issue or for estuaries in general. This can only be done for a specific scheme in a given estuary. For example, Kirby (1997) uses the essential character of the present physical and biological systems of the Severn Estuary to provide a platform to anticipate physical and biological changes arising from the proposed barrage in that estuary.

14.1 UNIQUE EFFECTS

Many of the effects of a tidal power development are essentially the same as those which would be experienced by construction of any major marine project. Dredge and fill, site access and traffic, and disruptions during construction, all fall into this category. Additionally, however, a tidal-electric generation facility will produce unique and important changes stemming from the nature of its operation.

Lands bordering the estuary or controlled basin are alternately inundated and exposed as the tides rise and fall. Closest to the extreme low-tide level are lands that, only occasionally, are exposed and drained. On the other hand, areas extending away from the shores of the estuary may be only rarely submerged during periods of highest high tides. Operation of a tidal power plant designed for ebb generation would permanently alter this natural condition because: (1) the basin level upstream of a tidal power barrage would normally not fall as low nor rise as high as under natural conditions; and, (2) the average basin level would be maintained above mean tide level. Mud flats bordering the basin would diminish in extent and marshland vegetation could encroach seaward. This would, in turn, change the habitat for wildfowl, shore birds, and miscellaneous species. The hydraulic characteristics both landward and seaward of the basin would also be altered, such as, currents, erosion processes, sediment deposition, and ice formation and movement. Figure 14.1 summarizes potential interactions between a tidal-electric development and elements of the environment.

Construction Activities and Plant Operation	The Environment[1]: Topography/bathymetry	Mineral Resources	Soil	Drainage and Surface Runoff	Gound Water	Tidal Regime	Fresh/Salt Water Interface	Physical Oceanography	Sedimentation/Erosion	Water Chemistry	Aquatic Ecosystems	Migrating Aquatic Species	Benthic Ecosystems	Intertidal Zone Productivity	Wetland Vegetation	Wetland Wildlife Habitat	Upland Vegetation	Upland Wildlife Habitat	Climate	Air Quality	Noise and Vibration	Ice Formation	Surface Water Hydrology
Construction activities																							
Site Development, Land-Based																							
Clearing, grading, surface excavation, building structures,material storage	•		•	•	•				•	•	•				•	•	•	•		•	•		•
Road rail spur construction	•		•	•	•				•	•							•	•		•	•		
Excavation for abuttments	•		•	•	•				•	•				•	•	•	•	•		•	•		
Material placement	•		•	•	•				•	•				•	•	•	•	•		•	•		
Operate land-based marine equipment	•		•	•	•			•			•		•	•	•	•	•	•			•	•	
Worker facilities and use	•		•	•	•					•						•	•	•					
Site Development, Marine																							
Pile driving	•		•					•	•		•	•	•	•	•	•		•			•	•	
Intertidal construction zone	•		•	•				•	•	•	•	•	•	•	•	•	•	•		•	•	•	
Dredging	•		•					•	•	•	•	•	•	•	•	•					•	•	
Mattress/dike placement	•		•					•	•	•	•	•	•	•	•	•					•		
Tug and barge operation	•									•	•	•								•	•	•	
Caisson storage and transport	•							•	•		•	•	•	•							•	•	
Caisson installation	•							•	•		•	•	•								•	•	
Stationary marine equipment	•							•	•	•	•	•	•	•							•	•	
Mechanical/electrical equipment installation																				•	•		
Site Accessibility																							
Road, rail transport of personnel, materials, or equipment			•														•	•		•	•		
Marine transport of personnel, materials, or equipment								•		•	•			•	•	•				•	•	•	
Remote Construction Facilities																							
Construction material source areas	•	•	•	•	•												•	•		•	•		
Dredge disposal sites, upland	•		•	•	•												•	•		•	•		
Dredge disposal sited, marine	•							•	•	•	•	•	•	•									
Operation of Permanent Facilities																							
Access and closure dike (presence)	•			•	•	•	•	•	•	•	•	•	•	•	•	•	•	•				•	•
Physical estuary barrier	•				•	•	•	•	•	•	•	•	•	•	•	•			•			•	•
Powerhouse and sluiceway (presence)						•	•	•				•	•									•	
Turbine operation						•	•	•	•	•		•									•	•	•
Sluiceway operation	•					•	•	•	•	•		•									•	•	•
Power facilities (presence)	•			•													•	•				•	
Switchyard operation																					•		
Drydock and dock facilities (presence)	•			•					•	•	•		•	•	•	•						•	
Long-term operation										•	•			•		•			•		•	•	
Impoundment (presence)	•			•	•	•	•	•	•	•	•	•	•	•	•	•			•			•	•
Water level fluctuation	•		•	•	•	•	•	•	•	•			•	•	•	•			•			•	•
Locks (presence)													•									•	
Operation								•	•			•	•								•	•	
Site access (presence)	•			•										•	•	•	•	•					•
Road, rail spur use																	•	•		•	•		
Marine use								•		•	•			•	•	•				•	•	•	
Worker facilities (presence)	•			•								•					•	•					
Use of workers					•							•									•		

1 •Indicates potential interaction betwen environmental element and plant component.

Figure 14.1 Potential interaction of a tidal power development and elements of the environment. (From Cook Inlet Tidal Power Study, Acres American Inc., 1981.)

The operation of a tidal-electric development will cause changes in water levels, water flow patterns and, velocities and sedimentation. Combined with the physical presence of the barrage, a development could exhibit the following impacts:

- *Impacts on Society* Agriculture, land drainage and flooding, water quality, opportunity for estuary crossing, employment, industry, amenities, recreational opportunities, ports and navigation, and shore protection
- *Impacts on the Environment* Birds, migratory fish, marginal wetlands, the estuarine ecosystem balance, and geological sites

The foregoing are not necessarily listed in order of importance since these impacts will have different values and influences in different estuaries. Most of the potential impacts arise from the altered tidal regime—that is, the range and pattern of water level fluctuations and the properties of tidal currents and would, therefore, generally be felt in relatively close proximity to the site. Figure 14.2 illustrates the relative changes in the basin level regime during ebb

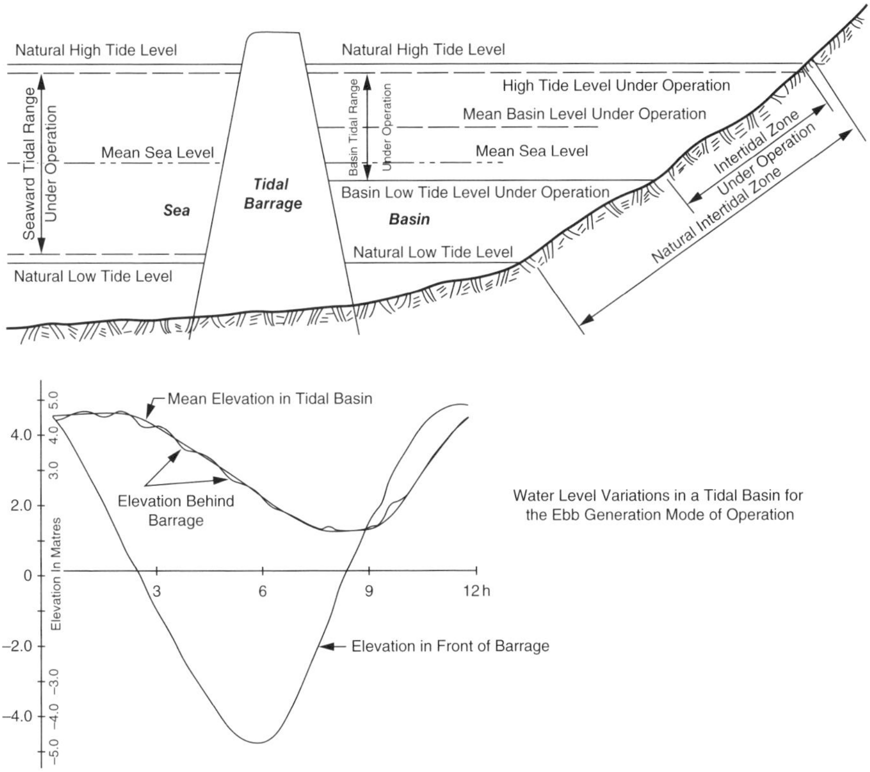

Figure 14.2 Relative position of tidal basin water levels under an ebb generation mode of operation.

generation, that is, single-effect operation. Figure 14.3 illustrates the relative regime levels under double-effect operation.

Indirectly affected would be wave climate, temperature and salinity characteristics, stratification, density currents, and ice formation. With regard to freshwater, both ground and surface water processes could be affected, such as flooding, draining of marsh lands, and saltwater intrusion. Changes in the foregoing characteristics could influence the climate, water quality, biological characteristics, and plants and animals.

14.1.1 Sea Level

Depending upon the scale of the development, its position relative to the resonance of the estuary, and the like, the tidal regime on coast lines relatively remote from the development could be affected. For example, although the effect of operation of a tidal-electric plant at the mouth of Cumberland Basin in the upper Bay of Fundy (Canada) would have no significant effect at Boston, Massachusetts, about 650 km south along the Atlantic coast, the larger development in Minas Basin (Canada) has been calculated to increase the range of the mean tide at Boston by 20 cm and of a large tide, depending upon the scale of the development, by 30 cm.

Changes in sea level occur for many reasons on different time and space scales. Tide gauges measure sea level variations in relation to a fixed benchmark and thus record *relative sea level* change due both to vertical land movements and *real* (eustatic) changes in ocean level. Vertical land movements result from various natural isostatic movements, sedimentation, tectonic processes, and even anthropogenic activities. Eustatic sea level is also affected by many factors. It is highly likely that global-mean sea level has been rising during the past century with most estimates lying in the range of 1.0 to 2.0 mm/yr, with the most likely causes related to climate change (Warrick and Oerlemans, 1990). Future changes in sea level due to global warming resulting

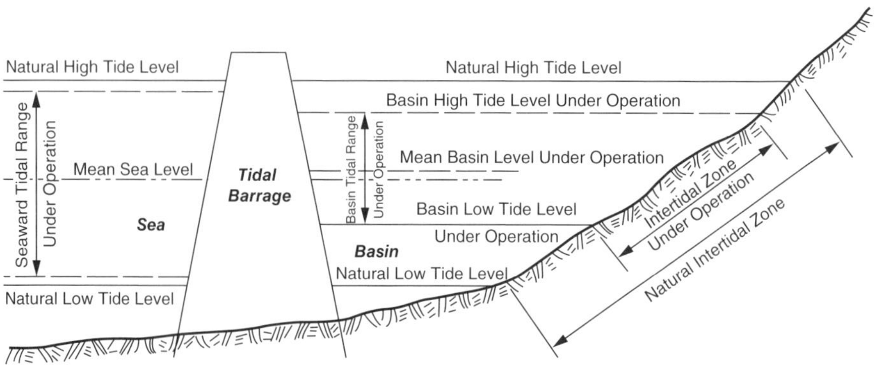

Figure 14.3 Relative position of tidal basin water levels for double-effect operation.

from current and future greenhouse gas concentrations have been estimated by the Intergovernmental Panel on Climate Change (1990). For the "business-as-usual" scenario, the best estimate of global mean sea level rise is 18 cm by the year 2030 and 44 cm by the year 2070. The panel also pointed out that even with substantial decreases in the emissions of major greenhouse gases, future increases in sea level are unavoidable—a sea level rise "commitment"—due to lags in the climate system. Such increases due to global warming should be of major concern to many low-lying coasts and overshadow possible increases due to the construction, and operation of a tidal-electric plant. In addition, possible sea level changes should also be reflected in the design of the tidal-electric plant.

14.1.2 Climate

The climate in the vicinity of a tidal power project could be affected by changes in the physical oceanographic regime such as the water temperature and, if applicable, the extent and duration of ice cover (Gates and O'Neill, 1977). Such changes could, in turn, influence the air temperature, cloud cover, and the amount and distribution of precipitation, both of which have an impact on the productivity of farmlands and forests.

14.1.3 Water Quality

Important chemical processes that may be affected include the cycling and availability of inorganic nutrients essential for the growth of aquatic organisms and the assimilative capacity of water for chemical pollutants. Potential water quality problems will depend on the estuary, which may already be receiving municipal, agricultural, and industrial wastes. The assimilation capacity of a basin is dependent upon many factors. One important factor relates to the estuary's flushing time, which will be reduced behind a tidal power barrage and, thus, could produce effluent disposal problems in the future. In addition, a new pollution problem might arise due to construction wastes or if the construction were to expose a geological formation that contained toxic trace materials that might enter the food chain. The reduced currents in the basin would result in reduced sediment mobility and, hence, alter the sediment–water interactions.

14.1.4 Geological Characteristics

Construction of a barrage and the operation of the tidal-electric development could alter the pattern of shoreline erosion and the sediment regime. In addition to having environmental implications, sediment transportation and deposition processes will likely have an important impact on the engineering aspects—design, construction, and operation—of a tidal power development (Amos, 1979; Amos and Asprey, 1979).

Siltation, resulting from structural interference with strong tidal flows that move large quantities of sediment, may be significantly changed when these flows are redirected. The two photographs in Fig. 14.4 were taken 10 years apart at the site of the Shepody River Control Dam in New Brunswick (refer to Fig. 16.8). Photo A was taken in 1956 following the construction of the dam and Photo B shows the accumulation of the large mud flats to its seaward side after 10 years. It is evident that in tidal near-shore regions characterized by highly silt-laden water, rapid and extensive sedimentation will occur when the tidal flows are obstructed. Observations on the effect of such structures has shown that siltation associated with intertidal construction is fairly rapid but decreases exponentially with time (Kestner, 1975). Postbarrier sedimentation patterns can be adequately forecast by mathematical models (Hydraulics Research Station, 1980; Greenberg and Amos, 1983; Nekrasov and Romanenkov, 1997).

14.1.5 Plants and Animals

The biological processes controlling the distribution, abundance, and productivity of living organisms are affected by the aforementioned physical, chemical, and geological processes. The estuarine environment may or may not already be highly stressed. A high-stress environment is one in which there is significant variation in parameters integral to the support of life forms. For example, salinities may vary dramatically from season to season as freshwater flows fluctuate, which may also cause substantive fluctuations in suspended sediment load and, in some regions, temperatures may vary seasonally by as much as 15°C. Such extreme conditions create hardships for organisms living in the water. If a high suspended sediment concentration were present, which would lower the quality of the water for growth of zooplankton and severely limit the penetration of light necessary for primary production of phytoplankton, the basis of the estuarine food chain, the result would be low biological activity in both the planktonic and benthic environments. The benthos is that portion of the estuary floor that is always submerged, and benthic organisms are the bottom dwellers of the estuary. Because they partially depend on organics derived from biological productivity of the water column, the amount of food sources would be dependent upon the degree of light penetration.

The foregoing factors could also stress the organisms in the intertidal region. In addition to temperature and salinity variations, the region is subjected to substantial tidal ranges (the reason for it being considered for a tidal-electric development) and waves and alternating currents that continuously inundate and drain the land at the waters edge and scour, erode, and deposit transitional sediments.

Inland of the mean high-tide line, the frequency and duration of tidal inundations decrease. These lowlands and marshlands are generally highly productive, evincing a great variety of vegetation types, tolerant to saltwater, which support extensive habitats for waterfowl and shorebirds. In summary,

(a)

(b)

Figure 14.4 Shepody River Control Dam. (Photographs illustrate sedimentation that took place in a 10-year period): (*a*) December 1956 and (*b*) November 1966. (Courtesy of Canada Department of Regional Economic Expansion.)

the biological response to a change in environmental regime may take several forms and may be formulated as follows:

1. Changes in the distribution of species within the estuary
2. Changes in the species composition of communities by losses of existing species, invasions of new ones, and alteration of the biological interaction between species
3. Changes in the local interaction between living organisms and geophysical processes through such actions as reworking of sediments by burrowing animals and accretion of sediments by salt marsh plants
4. Repercussions through the food chain resulting from changes in the species composition of communities that are preyed upon
5. Changes in the well-being of colonizing species, measurable in terms of growth rates, fecundity, reproductive success, and production of particular species

Three possible impacts on the terrestrial environment brought about by the construction and operation of a tidal-electric development would include changes in the availability of certain types of land (i.e., salt marshes, diked farmland, etc.), the productivity of farm and forest lands, and the availability of wildlife habitat.

14.2 AGRICULTURAL ASPECTS

The creation of a tidal basin may, depending upon its topographical features, impact adversely upon regional agricultural practices and crop production. Primary concerns are the control of water levels in the low-lying farmlands surrounding all potential tidal basin areas, and climatic changes capable of affecting crops over a much wider area. Peripheral concerns would include flooding and drainage aspects related to the modification of tide-induced water levels in the area. High-quality pasture and farmland may have been reclaimed from the estuary and now be protected by dikes that contain drainage sluices, also known as *aboiteaux*, to facilitate the discharge of surface runoff at times of low tide. In some locations dikes and aboiteaux may be protecting areas of commercial and residential development that have been permitted to encroach onto the floodplain.

Operation of a tidal development in the ebb generation mode would result, within the tidal basin or headpond area, in substantially increased low and mean tide levels, but slightly reduced high-tide water elevations (Fig. 14.2). This modified regime might significantly reduce the period during which draining of some low-lying farmlands could take place. However, the effects of any such installation would require individual determination and could require an increase in the capacity of drainage sluices. A corollary effect of increased mean water levels in the tidal basin areas could be the raising of groundwater tables in the surrounding lands. This could cause impaired drainage thereby resulting in reduced crop production.

The reduction in high-tide levels could have the beneficial effect of facilitating the more rapid release of impounded floodwaters during spring freshet periods. At the cost of some reduction in the energy production from the tidal power installation, the headpond level could be lowered to some extent to assist in mitigating a flood in estuarine reaches of tributary streams. Reduced high water elevations in the headpond area could lead to the reclamation of additional diked lands for agricultural use; reduced dike maintenance and decreased erosion in headpond shores would also be a consideration. While some salt marshlands may be eliminated initially, it is anticipated that deposition of sediment resulting from reduced tidal flows would provide a substrate for their reestablishment.

14.3 FLOODS AND DRAINAGE

If the tidal-electric development were operated in the ebb generation mode, the mean water levels within the tidal basin would be significantly higher by comparison with the natural situation. Furthermore, the development may also increase or decrease the tidal amplitudes seaward of the barrage, depending upon the size of the estuary that would be isolated and its location.

Although they may directly cause flooding, higher water levels seaward of the barrage resulting from increased tidal amplitudes could induce backwater flooding in some river estuaries. Higher backwater levels in such river estuaries could result in greater overbank discharges so that peak river flows, coincident with highest tides augmented by operation of a tidal-electric installation, could cause increased flooding.

Within the tidal basin, three changes may be significant. First, there could be a benefit through reduced flooding as a result of the lower high-tide levels both around the basin and within the estuaries of its tributary streams. Second, with a decrease of tidal exchanges, mixing energy in the headpond would be reduced. In some cases, a gradual decline in salinity could result. In northern latitudes, this would enhance ice formation. Lower high-water levels could also lead to accumulations of raft ice on intertidal flats at the mouths of tributary rivers, although the overall reduction of intertidal area, the lower production of ice due to reduced tides and the lack of sea ice transported into the enclosed basin areas would tend to counteract this effect. The overall result may be to reduce the ice problems in the tidal basin.

Third, there is concern regarding sediment deposition. Sediments transported by the strong tidal currents tend to settle at the mouths of incoming rivers thereby reducing hydraulic capacity and contributing to ice jamming. For example, this is thought to be the principal reason for periodic flooding at the mouth of the Salmon River at Truro, Nova Scotia, in the upper reaches of the Bay of Fundy (refer to Fig. 16.8).

The question of changes in sediment transport as a result of constructing a tidal barrage has a number of engineering ramifications. For example, the large, highly visible, modifications to channel bathymetry that have occurred in the Avon River estuary near Windsor and the Petitcodiac River at Moncton as a result of causeway construction, while not considered representative of the anticipated effects of a major tidal barrage, have nevertheless drawn attention to the need for very careful analysis of sediment dynamics in planning a tidal development (Amos, 1979). There are indications that, while sediments in the Bay of Fundy originate from both erosion of foreshores and drainage of inland areas, the large buildup of sediment at the mouths of several rivers is derived from seaward sources. It is important, therefore, that sediment source budgets be fully defined (McManus, 1979). If it were found that the principal source of sediment in an estuary is to seaward, it may be expected that a barrage would lead to reduced floodplain inundation.

14.4 BIRDS

Changes in regime within tidal basins would result in reductions in the areas of the intertidal zones. Populations of both waterfowl and shorebirds could be adversely affected. Indirect effects due to changes in the primary and secondary levels of biologic production within the intertidal zones would also be experienced.

While the loss of mudflat and marshland resources could be significant, land reclamation may have a greater impact on avifauna than is likely to result from a tidal power development. Experience elsewhere has also shown that shore birds are seldom food limited. They will crowd together on mudflats in response to a loss of habitat and show considerable flexibility in habitat selection (Longhurst, 1977).

There are three primary mechanisms by which the tidal plant could directly cause impacts to marshlands. These are due to: (1) interaction along the shores of the impounded basin; (2) interaction with the construction site, noise, activity, and equipment; and (3) interaction with an altered flow regime downstream of the dam. Of these three primary impacts, the potentially most significant would be the effect of the altered tidal regime on the stability and productivity of the marshland ecosystems within the tidal basin. Altered sedimentation patterns could result in eroded shorelines. A raised water level could result in a more saline groundwater table. Altered surface hydrology may affect filtering and transport of nutrients and organisms within the marsh. A loss of marsh area and a loss of vegetation types required for support of bird populations would diminish productivity and result in degradation of the waterfowl habitat. On the other hand, sedimentation could result in an enlargement of marshlands. Effects of changes in hydrology, inundation, and nutrient supplies could produce a habitat more attractive to waterfowl and other species. The interactions among the various factors affecting the wetland ecosystem must be understood before the effects of a changed regime can be predicted with reliability.

14.5 FISH AND FISHERIES

A tidal barrage and the operation of its turbines and sluices could pose a number of possible threats to fish stocks, either because of the barrier to passage or mortality during passage. Thus, fisheries concerns can be considered in three areas: passage of fish past a barrage, production in the headpond or tidal basin area, and effects on fisheries seaward of the barrage. The study of fish damage in turbines has related primarily to hydroelectric projects and, until recently, expected impacts have been gauged on the experimentation and experiences at hydroelectric stations (EPRI, 1987). However, research on fish mortality at the Annapolis Tidal Generating Station has indicated mortality rates somewhat lower than expected. Based on this result, Baker (1987) pointed out that nearly all conclusions about fish mortality through turbines has been based on research involving salmonids, whereas development of tidal power may involve the exposure of a wide range of anadromous and catadromous species to the hazards of turbine passage.

The magnitude of fish mortality varies between types of turbines and the head under which the turbines operate (Collins, 1984). Fish mortality through Kaplan turbines is significantly lower than that through Francis turbines; the former may have up to six blades and are used under low heads compared to Francis turbines, which may have more than twice as many blades so that the water passages through the Kaplan turbine are comparatively larger. Some fish species are intolerant to sudden pressure changes. Localized pressure differentials within the water passages of a turbine can be significantly different from the overall average pressure differentials. When the absolute pressure at any point in the water passing over the runner blades approaches the vapor pressure of water cavitation starts; small bubbles of water vapor will form in these low-pressure areas and, when carried into the main water flow at considerably higher pressure, will implode causing an intense impact to any adjacent metal parts, which will eventually be eroded—the phenomenon of cavitation. Cavitation can be practically eliminated by setting the turbine runner sufficiently below tailwater level to ensure that minimum pressures along the surface of the runner blade are well above the vapor pressure. Because cavitation is generally so localized within the turbine it is not likely to have a large impact on fish mortality.

Shearing action through the turbine may, however, be a significant source of mortality, particularly in tidal-electric plants. In such plants, the turbines operate at low head and low efficiencies at the beginning and end of each operating cycle (refer to Chapter 5) and tests have shown that shear damage increases as turbine efficiency decreases. It should be noted that with the use of asynchronous speeds, higher efficiencies are attainable at the very low heads available at either end of the operating cycle.

A number of rivers flowing into the tidal basin may support spawning runs of anadromous fish so that a tidal barrage would create a barrier to the free passage of fish in either direction past the plant location. Anadromous fish, however, swim in the surface waters, and the maintenance of runs may necessitate special arrangements. It is likely that submerged gate and turbine openings would be at least 2 m below water level and could be greater than 6 m from the surface. It is not known whether anadromous species will sound to such depths to enter the tidal basin. Since this is considered to be unlikely, it will probably be necessary, where anadromous fish are indigenous to the estuary, to include one or more surface gates that could be moved in phase with the constantly changing water levels to maintain the fish runs when they occur. Potential problems relating to the inward migration of spawning fish arise from possible delay, disorientation, and failure to reach spawning grounds (Semple, 1984).

With respect to the pelagic species, the effect is likely to be minimal. Low-speed turbines do not create conditions likely to injure the fish and mechanical abrasion should not be significant. There will be substantial periods during each tidal cycle when fish can pass in an upstream direction through the sluiceway gates and downstream through the turbines. However, in some cases, such as for the shad along the Atlantic coast of North America, it has been suggested that the migration and swimming capability of juvenile shad could be adversely affected and result in a lower survival rate than had been estimated (Leggett, 1977). This underlines the need to redirect impact assessment studies as basic environmental studies proceed.

In those estuaries visited or inhabited by species that may pass in and out of the estuary a number of times, such multiple passages would significantly increase any single-passage mortality figures. In such cases, means must be provided to divert fish from the turbine route to alternative routes past the barrage, such as by sluices or fishways. Such methods are employed at hydroelectric plants but may not be effective for all species frequenting or indigenous to the estuary. Behavioral devices, such as lights, sound, and so forth, combined with unobstructed water passages, may provide a solution but further research will be necessary.

A tidal-electric plant may enhance marine culture in its basin. Rétière et al. (1997) point out that the flora and fauna in La Rance (France) basin was, as of 1996, rich and diversified. In addition, it was also reported that there are now 110 species of marine worms, 47 species of shellfish (compared with 44 before construction of the barrage) and 70 species of fish, the same number as prior to beginning of construction in 1963.

14.6 RECREATION AND TOURISM

Various features of the shores of the tidal basin and the tidal-electric plant itself could combine to form the basis for increased interest in tourism and recreational activities. As an example, La Rance development has now been in operation for more than 30 years (refer to Chapter 15) and has attracted up to 400,000 visitors per year, placing it in the first rank of French industrial tourist sites (Barreau, 1997).

Changes in tidal amplitude may affect beach erosion in some areas that would require monitoring and remedial action; habitat modifications could affect established recreational patterns. A reduction in the tidal range within the tidal basin area could alter the thermal regime, thereby improving conditions for recreational use.

The barrage would provide the basis for a road to cross the estuary, which would give a substantial impetus to development in the region surrounding the basin and improve traffic flows.

14.7 TRANSPORTATION

A major effect of a barrage would be to enclose the ports in the estuary behind a barrier. The barrage would affect ports and navigation by requiring ships to pass through locks in the barrage, by changing the water levels and, possibly, by modifying the pattern of sedimentation. The latter two factors might also influence ports immediately seaward of the barrage.

Land transportation could be materially enhanced since the barrage would provide the means for a highway crossing, eliminating the much longer route around the estuary. This could give a substantial impetus to development in the region surrounding the basin and improve traffic flows. For example, La Rance barrage provides a roadway across the mouth of the estuary, which is now traversed by up to 50,000 vehicles on a summer day (Barreau, 1997).

14.8 SUMMARY

In summary, changes in water levels: (1) determlne the energy that could be extracted from a barrage; (2) influence navigation and the drainage of low-lying land; (3) could lead to major recreational and amenity benefits above a barrage; and (4) would alter the exposure of the present foreshore and inter-tidal banks, which would affect the habitat of wading birds and waterfowl. It is, therefore, important to be able to know how accurately the new water levels can be predicted.

Changes in water velocity and flow patterns would affect the sediment load carried by the water and the sedimentation patterns. These in turn could have important implications for navigation and the natural environment; the dispersion of any pollutants in the controlled estuary would change, which could, in turn, have industrial impacts as well as environmental effects.

The pattern of water movement would be affected in several ways: (1) The flow pattern would change as construction of the barrage progressed;

(2) water velocities would generally be significantly reduced behind the completed barrage due to the reduction in tidal range; (3) similar reductions in water velocities seaward of the barrage would also occur but diminishing with distance from the barrage; and (4) the flow pattern would be modified because the barrage would constrain flow to the turbine and sluice passageways at predetermined times. For example, in an ebb generation scheme, about three-quarters of the flood tide flowing into the basin would pass through the sluices with the remainder entering through the turbines, which would idle in reverse, but all of the outflow would pass through the turbines. Since the turbines are likely to be in a single block in the deepest water, the outflow would be a coherent strong flow. The sluices would be in the next deepest water in two or more blocks so the inflow would be made up of several streams spread out across the estuary, diffusing within a few kilometers from the barrage.

Changes in the pattern of sediment transport, deposition, and erosion in the estuary could result in a wide range of possible impacts, such as on the performance of the works should sediments be scoured or deposited in or near the sluice and turbine waterways or erode the turbine runners and other parts of the works; on the maintenance of navigation channels; and on marine biology by reducing a turbid water condition with a consequent increase in light penetration and the several effects this could have.

The various impacts and their causes are to a large extent interrelated and involve many complex issues. An estuary would be truly unique if sufficient information already existed to enable evaluations to be made of the environmental alteration, the effects on natural processes and the natural resource availability. A major environmental field data collection and analysis program is, therefore, necessary in order to define environmental impacts with some degree of precision, although in the field of environmental evaluations there will always remain a number of uncertainties. In fact, the conclusions reached on environmental impact will inevitably be less certain than those reached on technical feasibility or on economic feasibility. Moreover, it will not be possible, nor indeed desirable, to quantify some impacts in economic terms.

Until such data and information can be amassed, the feasibility study must rely on the existing, limited database to identify, at least, any ecological, social, and industrial impacts that might rule out any or all proposed schemes or suggest the types of modification and postulate initial conclusions on various potential impacts, beneficial and adverse, on the basis of expert analysis of the best information available. A prefeasibility study should outline the studies necessary to fill gaps in the existing data and, perhaps, initiate those with a long lead time.

Any tidal-electric generation facility should be considered as only one prospective component of the future generation mix of the regional utility system. Thus, the assessment of environmental implications of tidal projects should be placed in perspective through comparison of system impacts with and without a tidal-electric generation facility (Waller, 1972). In so doing, judgments of nonquantifiable aspects may be approached on a comparative basis and lead to a better appreciation of the relative importance of potential impacts, both beneficial and adverse.

The reader is also referred to Chapter 15 and the summaries of the environmental effects resulting from the construction and operation of La Rance and Annapolis tidal-electric power plants.

CHAPTER 15

EXISTING TIDAL-ELECTRIC DEVELOPMENTS

As noted in Chapter 1, the major technological problems formerly associated with large-scale power projects to exploit tidal energy have now been resolved, and their output has been shown in several instances to be competitive with alternative energy sources. Moreover, this energy source is renewable, pollution free, relatively inflation free, and completely predictable. Estimates have placed the total capacity of all potential sites in the order of 1,000,000 MW with an expected output of, perhaps, 3000 TWh (Bernstein, 1979). The obvious question is: Why is this resource not now being exploited to its fullest extent? The answer is complex, but it undoubtedly is a function of government policies and bureaucratic reluctance to undertake a project that, although technically feasible, has no precedents in the field of successful "giant" projects as well as the high initial investment required compared to the less capital-intensive, fossil-fired, thermal power stations.

Subsequent to World War II, interest in the conversion of tidal energy to electrical energy began to gather momentum. Although a number of tidal-electric plants have been built during the past quarter-century, most have been miniscale developments (<1 MW) except for the Jiangxia plant (3 MW), which may be classified as small scale, and two larger-scale projects: La Rance (France) plant of 240 MW, which was commissioned in 1966, and the Annapolis (Canada) plant of 20 MW, commissioned in 1984. Future developments will profit not only from substantive technological advances, for example, in construction in the marine environment, in turbogenerating

Elements of Tidal-Electric Engineering. By Robert H. Clark

machinery, and in mathematical simulation techniques, but also from the information and experience gained in construction and operation of the modern tidal-electric stations, described briefly in this chapter.

15.1 LA RANCE TIDAL-ELECTRIC PLANT

Conceived as a large prototype from which to gain experience for possible development of much larger schemes to harness the tidal energy potential along the Brittany Coast of France, the Rance tidal-electric plant was built between 1961 and 1967 with first energy being produced in 1966. The location and plan of the scheme is shown in Fig. 15.1. The following is a summary description of this unique project based on Cotillon (1979).

La Rance estuary is about 750 m wide at the project site. The barrage consists of a shiplock, a powerhouse, a short rockfill section, and a sluice structure of 115 m in length, with six fixed-roller gates (15 m wide × 10 m high) capable of passing the flow at mean tides (9600 m^3/s) under a 5 m head. Sectional elevations of the powerhouse, the rockfill dike, and the sluiceway are shown in Fig. 15.2. These components were constructed in the dry behind cofferdams, as illustrated by Fig. 8.1. The construction of the project involved four principal stages: (1) 1961—execution of cofferdam enclosures for construction of the navigation lock and sluices; (2) 1962–1963—construction and equiping of the navigation lock and of the sluices; execution of the central cofferdam enclosure to enable construction of the powerhouse; (3) 1964–65—erection of the power station and of the rockfill dike; assembly of the first generating groups and other electromechanical equipment; and (4) 1966–1967—flooding of the central enclosure and dismantling of the cofferdams; completion of installation of the electromechanical equipment and setting it into operation. Figure 15.3 shows the completed structure, as of November 1966, with almost all of the cofferdam dismantled.

Figure 15–4*a* is a view of the water passage from upstream of the bulb that contains the generator while, Fig. 15.4*b* shows the runner from the draft tube side. The turbogenerator units are double regulated, reversible bulb turbines capable of turbining or pumping in both directions as well as sluicing in both directions. This flexibility was built into the plant, although only a single-basin scheme, in order to provlde as much information and experience as possible about the operation of tidal-electric plants. Twenty-four units, with 5.35-m runner diameter and rated head of 5.65 m, were installed, each directly coupled to a 10-MW generator-motor. The maximum head is 11 m and the minimum, 3 m. The 24-unit installation does not represent the optimum installation for this estuary, which should be in the order of 40 units (Gibrat, 1976) but, rather, it reflected limitations imposed by the available capital.

The sluiceways allow complete emptying of the basin at the end of the direct generation period or complete filling at the end of reverse generation, when the difference in head between basin and sea is smaller than the minimum

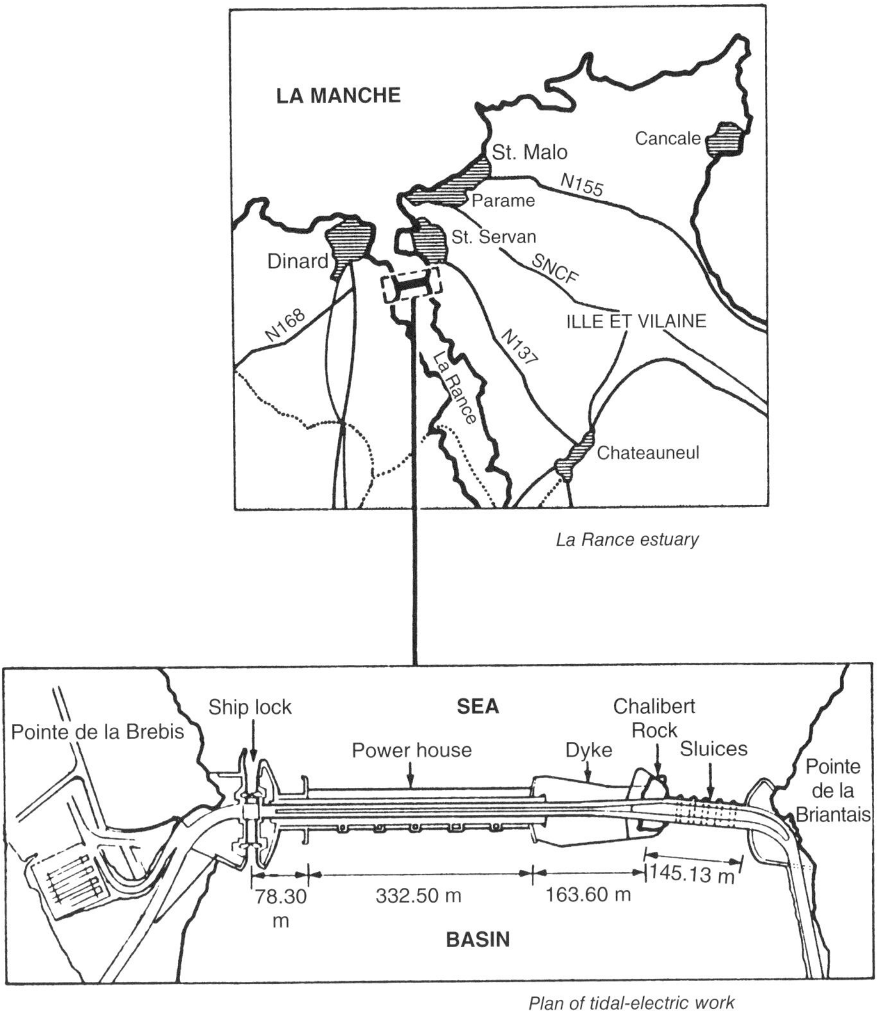

Scheme details

Estuary width:	*750 m*
Basin area:	*22 km*^{2}
Mean tide:	*8.5 m*
Installed generating power:	*240 MW*
Turbogenerators:	*24 × 10 MW bulb-type Kaplan turbines with reverse-flow and pumping capability*
Turbine runner diameter:	*5.35 m*
Rated head:	*5.65 m*
Maximum head:	*11 m*
Minimum head:	*3 m*

Figure 15.1 General layout of La Rance tidal-electric station.

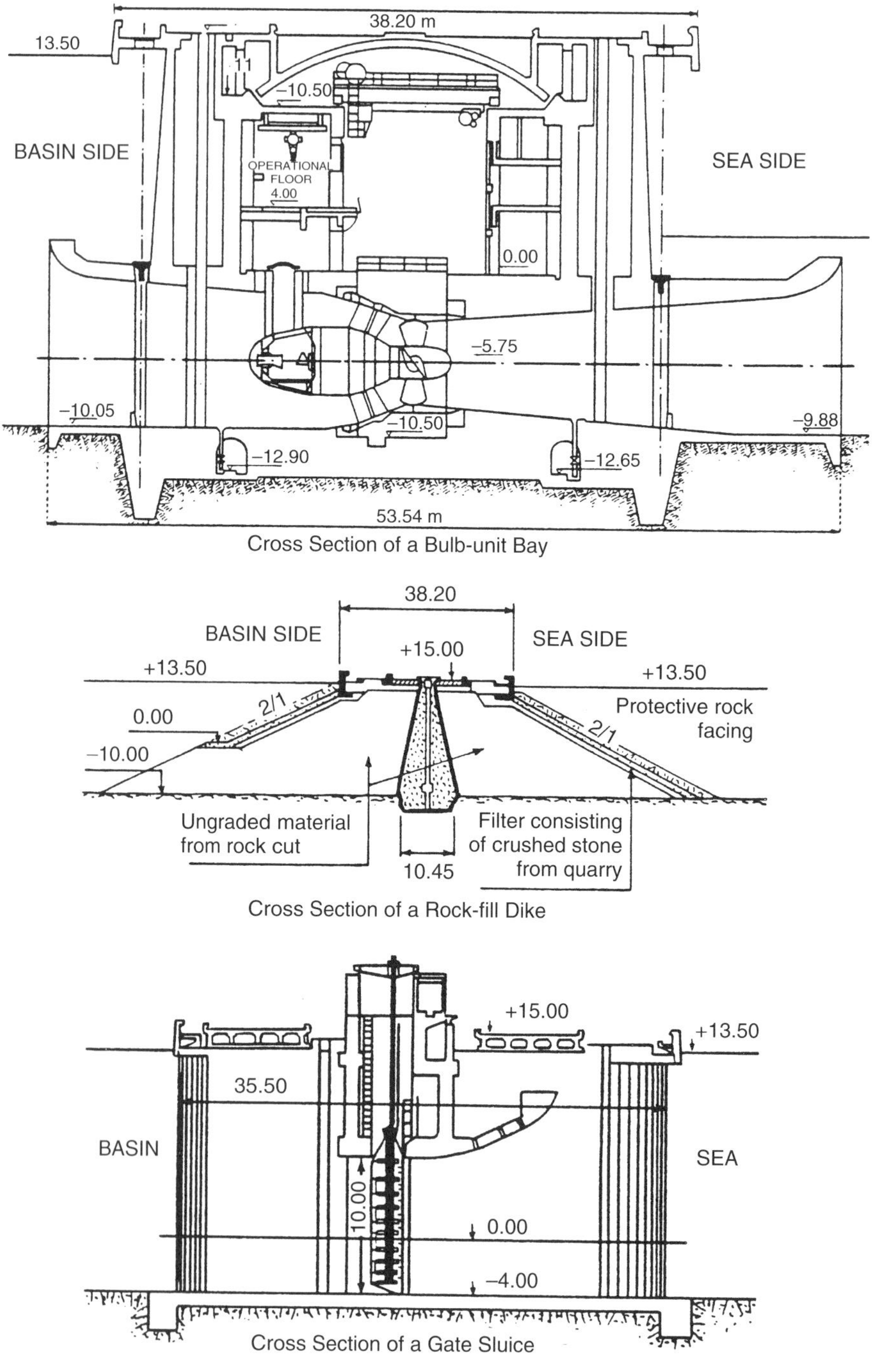

Figure 15.2 Sectional elevations of component structures. (From Cotillon, 1979.)

Figure 15.3 La Rance tidal-electric station. November, 1966, cofferdam almost dismantled. (Courtesy of H. Baranger et Cie.)

(*a*) (*b*)

Figure 15.4 Views of La Rance bulb turbogenerator: (*a*) View from basin side. (Courtesy of H. Baranger et Cie.) (*b*) View of runner from sea side. (Courtesy Doucet-Dinard.)

working head for driving the turbines, that is, 1.2 m outward when emptying the basin or 1.6 m inward when filling. The sluice section is also used for total filling of the basin in the single-effect, ebb flow mode of operation. As already noted, the turbine units also operate as orifices to assist the sluices.

15.1.1 Tides

The average tide is 8.5 m and is of the semidiurnal type. Figure 15.5 illustrates the variations in tidal range. There is no significant seasonal effect throughout the year but a 2-week recurring cycle can be identified, comprising roughly 1 week with spring tides between 9 and 12 m, and 1 week with neap tides between 5 and 9 m. The ratio between the ranges for ordinary spring and neap tides is of the order of 2.5 and is about 4 between the extremes. Characteristic data for the tides at the La Rance site are provided in Table 15.1.

15.1.2 Operational Experience

A critical period for the plant was 1975–1982 (Electricité de France, 1992). In January 1975, following verification of a reduction in the air gap between the rotor and stator of one of the machines, it was concluded that the stresses

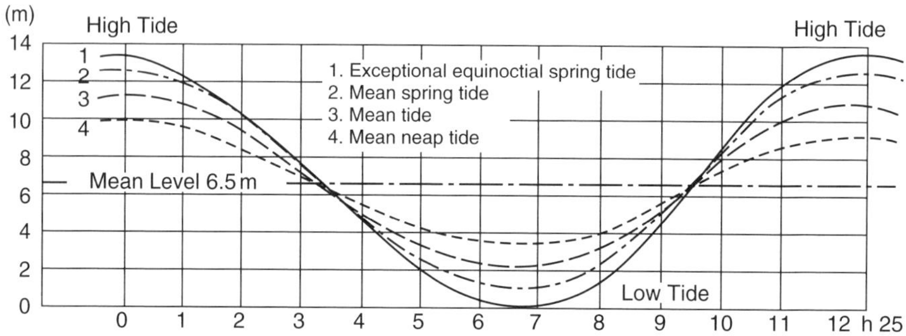

Figure 15.5 Variations in tidal range at La Rance.

TABLE 15.1 Tidal Characteristics at La Rance Site

	Tidal Coefficient	Tidal Range (m)	Vol. Water Stored ($10^6 m^3$)	Discharge (at flood) (m^3/s)
Equinoctial spring tide	120	13.5	184	18,000
Ordinary spring tide	93	10.9	150	12,900
Mean tide	70	8.5	110	9,400
Ordinary neap tide	45	5.0	65	5,500
Minimum neap tide	35	3.5		4,000
Mean water level		6.7		

linked to the asynchronous startups for pumping had been underestimated so that every stator had to be rebuilt, and this task took until May 1982. It was carried out with as little loss of power and energy output as possible; in fact, the station sustained about 80 percent of its designed output. Otherwise, problems have been of a minor nature similar to those one might expect in an electricity generating plant. Electrical spark erosion of the rotor poles had been observed in 1968 so that all 1536 poles eventually were dismantled for modification, and this work was completed during the 1975–1982 period (Salvi, 1997). The opportunity was also taken to examine the turbines, and they were found to have suffered no notable oxidation of their components after about 130,000 h of submergence and 90,000 h in operation. The overall plant availability, since 1982, has been about 97 percent. By carrying out routine maintenance during neap tides, loss of energy is kept to a minimum so that energy availability has been about 96 percent. The total net electricity output has been about 0.5 TWH/year. The flexibility of the double-effect operation enables the barrage to respond occasionally to unscheduled requests for energy.

The selected rated head at La Rance is 5.65 m and the design flow is 6600 m^3/s, which correspond to a tide about halfway between an ordinary neap tide and the mean tide. Although the rated head results from an economic choice of the design head of the machines and the design flow of the power station based on the characteristics of the tide and basin site, it will be noted that the rated head is lower than the available head for spring tides. The rated head is the minimum head at which the rated output (maximum output) can be reached. Taking account of the 2-week tidal cycle, two-way or double-effect operation involving a reduced active head is attractive during 3 or 4 days of that cycle, and single-effect operation is satisfactory during the remaining period. When the actual head is less than the rated head, the turbines are operated between maximum output and the output corresponding to maximum efficiency in order to have a large enough discharge so that the flow from the basin will produce the required energy. A further factor that had to be taken into account due to the variation in ranges was the difference in output between a day of spring tide and one of neap tide. For example, for tidal ranges of 11 and 5 m, the ratio of outputs would be about 4 but, without pumping, it would be in the order of 5 since, for the spring tide, the head would be about 3 times greater and the generation period, twice as long.

Figure 15.6 illustrates the operational modes used, depending upon the magnitude of the tidal range. Generation in the single-effect mode, from the estuary to the sea (direct turbining), is illustrated in Fig. 15.6*a*. Every effort is made to store the greatest quantity of water in the estuary so that overfilling of the estuary is attractive, particularly during the neap tides, and direct pumping is applied, as illustrated. Generation under the higher tides is generally in the double-effect mode as illustrated in Fig. 15.6*b*. It will be noted from Table 15.2 that direct turbining represents almost 60 percent of plant operating time while direct pumping represents 18 to 20 percent. The two reverse functions, reverse turbining and reverse pumping, are very little used.

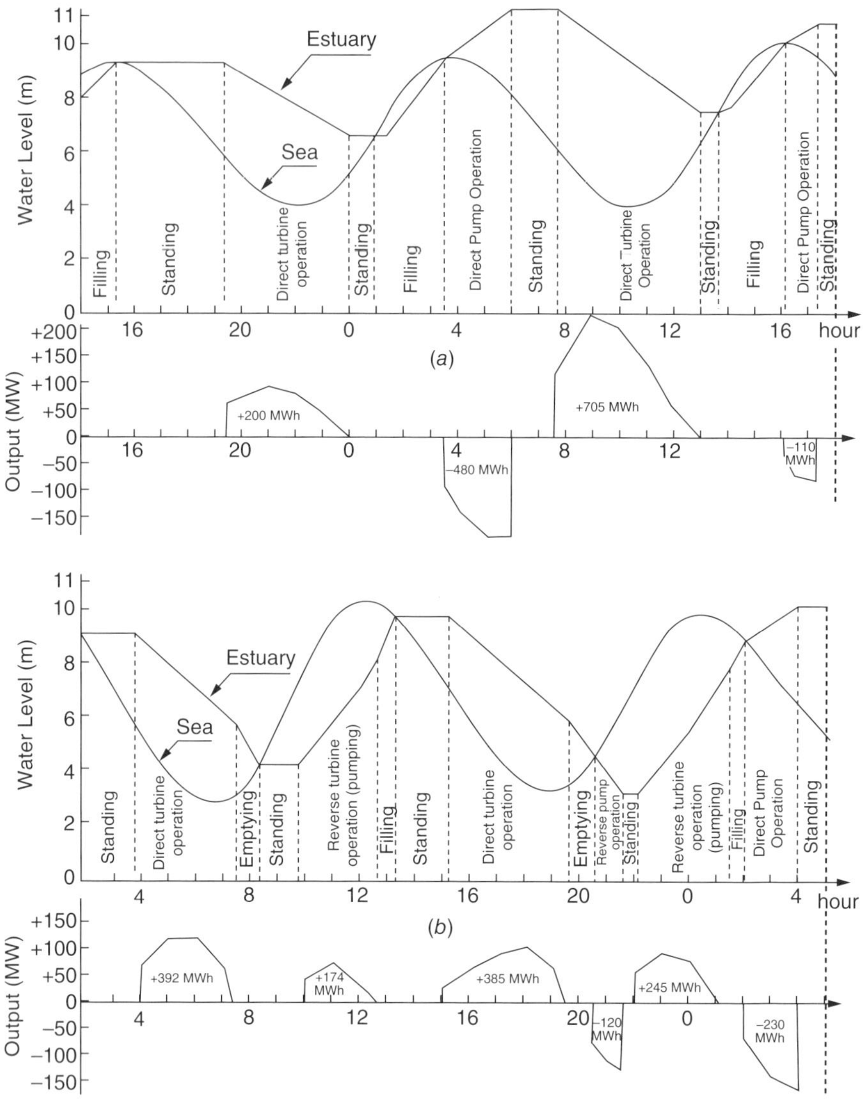

Figure 15.6 Operational mode and output for large and small tidal ranges. (*a*) Single-effect operation and (*b*) double-effect operation. (From Electricité de France, 1992.)

The turbines are used as orifices for about 20 percent of the operating time of the plant to complement the opening of the sluices during filling or emptying of the estuary basin. The distributor and runner blades are locked fully open for maximum flow, but the speed of the unit, under the heads then prevailing, is less than its nominal speed.

It was found that for tidal ranges greater than about 11 m, the energy output is significantly greater under the double-effect mode. Moreover, the

TABLE 15.2 Operation Summary, 1980–1990

	Production GWH		Operation Mode as Percentage of Total Operating Period					
Year	Gross	Net	D_T	R_T	D_P	R_P	O	Total
1980[a]	503	485	64	5	3	0.1	28	100
1981[a]	570	500	61	2	16	0.0	21	100
1982[a]	607	511	59	2	18	0.0	21	100
1983	610	503	57	6	17	0.1	20	100
1984	609	494	58	3	18	0.0	21	100
1985	612	485	58	5	19	0.0	18	100
1986	595	487	58	3	20	0.0	19	100
1987	576	479	57	6	19	0.0	18	100
1988	575	486	56	6	20	0.0	18	100
1989	566	495	58	6	20	0.0	18	100
1990	579	497	56	5	18	0.0	21	100

Source: Electricitié de France, (1992).
[a] Major normal maintenance work: 1975–1982.
Net production = gross − auxiliaries − pumping; consumption by auxiliaries = 8 GWH.
D_T, direct turbining (estuary to sea); R_T, reverse turbining (sea to estuary);
D_P, direct pumping (sea to estuary); R_P, reverse pumping (sea to estuary); O, orifice mode.

limited output for heads greater than the rated head deprives pumping of any significant attraction beyond a given range, and single-effect operation generates less energy than double effect. Based on operating experience, Cotillon (1979) suggested the following guidelines for the operation of a single-basin tidal project, for example, La Rance, should be as follows: (1) for spring tides, that is, tidal ranges greater than 11 m, double-effect generation is of interest because of the limited output of the machines; (2) for tides ranging between 7 and 11 m (specific figures apply to La Rance only since the observations are dependent on site and machine characteristics), either reverse generation or appropriate direct pumping may be considered, although in the former case there is a more continuous supply of power but a loss of energy; (3) for neap tides, that is, under 7 m, direct pumping has to be considered from the point of view of profit in terms of energy quantity or energy value; and (4) reverse pumping should be associated with reverse generation, whenever profitable.

It is interesting to note that after about 12 years of operation, the program to determine the optimal operation of the plant's 24 bulb units and 6 gates (Merlin et al., 1980) had to be revised in order to give it a higher degree of adaptability to the changing environment as well as more favorable characteristics of utilization. The revised operational model fulfills correctly all the necessary functions (Charon, 1997).

When generation starts, a turbine can be brought up to synchronous speed in less than a minute. Thus, all turbines could be brought on line in less than 2 min. However, care has to be exercised over the rate at which the turbines and gates are opened and closed in order to avoid waves traveling up and down the estuary.

A comprehensive impressed-current cathodic protection system was incorporated into the scheme. This system uses about 20 kW and has been successful in preventing corrosion of the stainless steel runner blades and removable draft tube adjacent to the runner as well as of the normal steel used for the bulb, the draft tube liner, and other submerged metal parts. Although a sophisticated paint system had also been used on the normal steel components, this would not now be considered necessary in view of the success of the cathodic protection (Lambert, 1997). In addition, to test the reaction of different materials to seawater corrosion, 12 of the turbines were fitted with stainless steel runner blades and 12 with aluminum bronze blades. An inspection, by the author, of the blades of several of the turbines in 1979, 13 years after commissioning, showed them to be in excellent condition, with even large areas of the blades still covered with paint.

15.1.3 Environmental Aspects

The total closing of the estuary between 1963 and 1966 during the construction of the plant caused the almost complete disappearance of the marine flora and fauna as a result of the consequent salinity fluctuations, heavy sedimentation, and accumulation of organic matter in the basin. When tidal flow resumed and the power plant commenced operation, the rythm and range of the tides were not what they had been. The wetted area had become smaller and the shape of the estuary had been affected slightly by the strong, separate currents through the sluice gates and through the turbines. Following commissioning of the plant and restocking of the estuary, aquatic life flourished again in the tidal basin and a new biological equilibrium was reached, though not the same equilibrium as had prevailed before the project began. To reach this stage required 10 to 15 years, or more. Strong productivity, the complex structure of the trophic system, and the passage through the turbines without excessive mortality of the migratory and nonmigratory species, testify to the current, normal, functional relations in the heart of La Rance ecosystems, as well as between the estuarial and coastal waters (Rétière et al., 1997). This biological equilibrium depends on the regularity of the power station's operation; sudden variations of the environmental factors can have drastic biological repercussions. Rétière et al. (1997) concluded that the success of a tidal-electric installation depends on the balance maintained between the restrictions on the power plant and their implications for basin ecosystems. The estuary again plays a nursery role for underwater creatures and remains a substantive home for birds (Rétière, 1984). The existence of the barrage has not accelerated the silting up of the estuary but may have led to sedimentary redistribution (Electricité de France, 1992).

15.1.4 Summary

The presence of the lock in the project has enabled pleasure boating to be developed in the estuary. The tourism economy that had already existed was greatly reinforced so that the number of visitors to the site is now on the order of 400,000 annually. From the environmental point of view, the presence of the barrage has produced a profound, but not negative, change in the ecosystem. Figure 15.7 is an aerial view of the La Rance tidal-electric station and ship lock looking toward the English Channel. Cotillon (1979) has stated that La Rance is not only a source of valuable lessons but also an economically attractive power station since the standard cost of tidal energy generated in 1976 was equal to the average cost of a nuclear kilowatt hour and also equal to the average system thermal kilowatt hour cost.

In summary, after 30 years of service, La Rance tidal-electric station has lost none of its ability to extract energy from the tides. Despite the very severe operating conditions to which they were submitted, the bulb turbines have performed well and will be ultimately renewed as originally designed (Bosc, 1997).

Figure 15.7 Aerial view of La Rance tidal-electric station. (Courtesy of Photothèque EDF: M. Brigaud.)

15.2 JIANGXIA TIDAL-ELECTRIC PLANT

China began to study the exploitation of tidal-electric power in the mid-1950s, and a 40-kW single-effect tidal-electric plant was completed in 1959 in Zhejiang Province. It has now been enlarged to give it a 200-kW capacity. Another single-effect plant, with a capacity of 155kW, was built on Jingang Creek in Shandong Provice in 1970. It has three 55-kW turbogenerators all made by a local machinery plant. This small station, jointly run by nine villages in a rural peoples' commune, supplies cheap power for farms and lighting for peasants' homes. About 280 pumping stations driven by tidal energy have been built along the coast of Guangdong Province since 1958 (*Modern Power Systems*, 1981).

In the 1970s there were 6 tidal plants with 19 turbines, aggregating more than 1900kW, installed successively in Zheijiang, Jiangsu, and Shandong provinces. The maximum heads at the sites vary from 3.5 to 7.8m and the installed unit capacities range between 75 and 160kW (Zhu, 1980).

The first bulb-turbine unit of 2.5m diameter in the Jiangxia tidal-electric plant in Wenlin County, Zhejiang Province (Fig. 15.8) was commissioned in April 1980. Although the proposed ultimate design capacity was to be 3000kW from six bulb-turbine sets, the current installed capacity (SHP News, 1997) is 3200kW from five bulb-turbine sets made up of one unit of 500-kW capacity, one unit of 600-kW capacity, and three bulb units of 700kW each. The proposed sixth unit, not yet installed, is to be a 700kW Straflo turbine. The

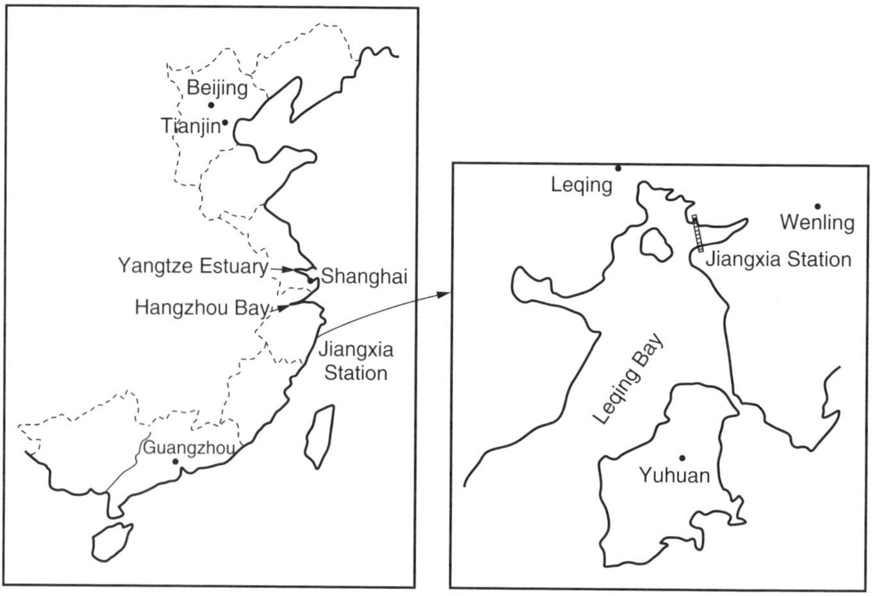

Figure 15.8 Location of Jiangxia tidal-electric power station.

Figure 15.9 View of Jiangxia tidal-electric power station from the ocean side. (Courtesy of Zhu Xiaozhang.)

increased capacity of the bulb-units was made possible by improvements in the design and manufacture of the turbines (SHP News, 1985). The maximum range of the tide in the Jiangxia estuary is 8.39 m. The plant is operated in the double-effect mode. The Jiangxia station feeds a small grid supplying villages 20 km distance through a 35-kV transmission system. Figure 15.9 shows a view of the project from the ocean side.

15.3 KISLAYA GUBA EXPERIMENTAL PLANT

The small Kislaya Guba experimental tidal-electric plant was put into operation in late 1968, with one unit of 400-kW capacity, at the mouth of that small bay, 100 km north of Murmansk on the shores of the Barents Sea (Bernstein, 1972), about 2 years after commissioning of La Rance station. The location is shown in Fig. 15.10. The USSR has considerable tidal enery potential, amounting to an estimated one-tenth of the developable tidal energy resources in the world. However, these sites are located on the coasts of the Barents Sea, the White Sea, and the Sea of Okhotsk, in uninhabited regions with a rigorous climate.

It was decided in 1962 to construct Kislaya Guba station as an experimental plant to verify the application, under the particular physical and climatic conditions of the region, of the float-in-place method or use of caissons (Fig. 8.4), at that time a relatively immature technology. The application of this method under USSR conditions, where the best sites for building tidal-electric stations are located on rocky coastlines with difficult land approaches and a bleak climate, seemed to offer considerable advantages, particularly since the station or floating components could be built in the favorable conditions of a seacoast industrial center, towed in a completed state (with equipment already assembled) to the site and sunk onto prepared foundations. Since the solution to problems that arose during construction of this experimental plant could be of value for hydraulic construction in general, and under the rigors of the far

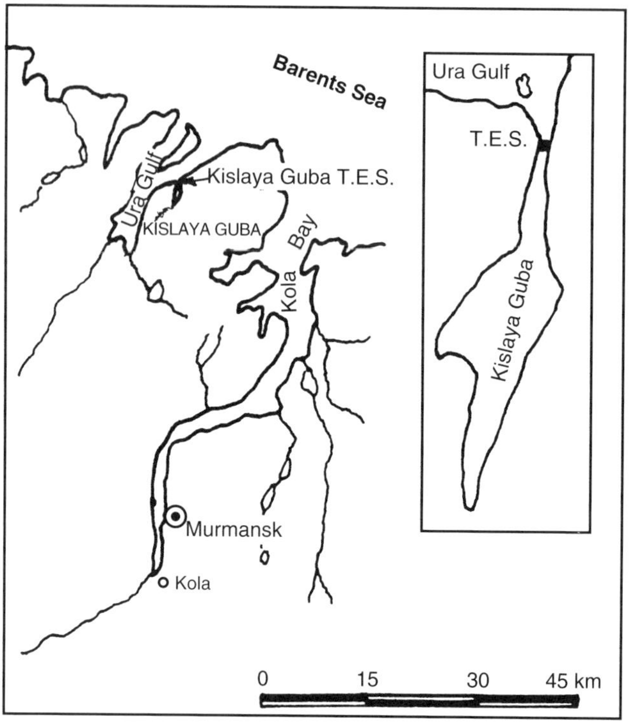

Figure 15.10 Location of Kislaya Guba tidal-electric station.

northern regions in particular, a broad program of comprehensive investigations was an integral part of the project.

The particular location of this scheme was selected primarily to require minimum work to close off the bay, which would become the tidal basin for the station. The neck of the sea connecting the bay to the Barents Sea is only about 50 m wide. Although the tide at Kislaya Guba has a range between 1.3 and 3.9 m, which would not make it an economic development, the site is relatively close to Murmansk which could benefit from its output. Moreover, with the relatively low range, it was possible to carry out investigations of the power characteristics of the installation under the most complicated conditions.

In the natural state, the tide filled the 1.14 km^2 bay through the narrow neck within 6 h 12 min, with the flow producing high speeds up to 3 m/s attaining and values of 300 m^3/s. The depths at the site were about 4 to 5 m and the bottom was of unconsolidated marine sediments 5 to 7 m thick. In the preparation of the site to receive the caisson, the deposits were excavated to a depth of 5 m by a floating crane (lifting capacity of 10 tons) equipped with a set of grab buckets. The total volume of excavated material was 25,000 m^3. However, the site preparation ran into a number of difficulties and its costs proved to be high. Techniques developed in connection with the Delta Plan Works could be significantly more cost effective.

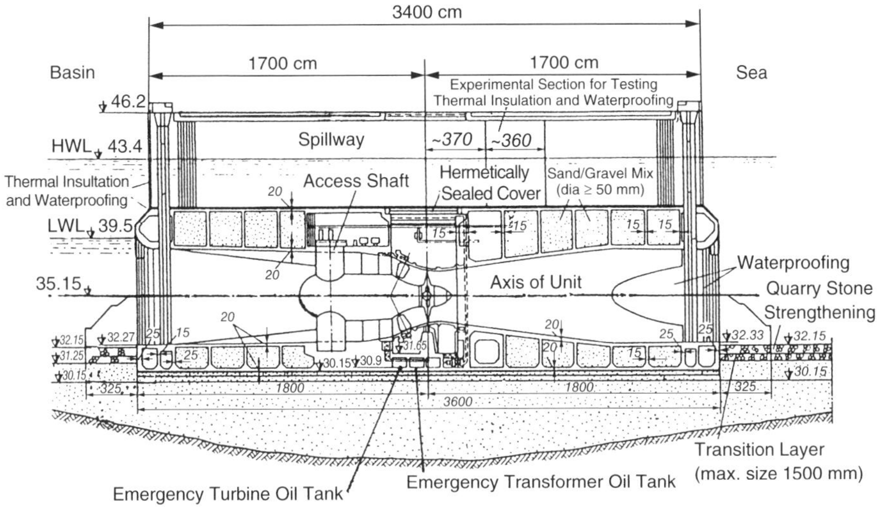

Figure 15.11 Sectional elevation of powerhouse caisson.

Figure 15.12 View of Kislaya Guba tidal-electric station from the sea side.

A sectional elevation of the powerhouse is shown in Fig. 15.11. The second unit was never installed. Figure 15.12 is a photograph of the Kislaya Guba tidal-electric station.

In 1974, Bernstein indicated that the experimental station had yielded much valuable information regarding construction and placement techniques, including foundation preparation and concrete durability under harsh climatic conditions. Unfortunately, very few, if any, environmental studies were conducted during the early period of operation. Also, from about 1974 until 1983 the plant was operated sporadically so that, for long periods of time, there was little or no water exchange between the bay and the sea, resulting in a dramatic collapse of the ecosystems in the bay. However, begining in 1984, the experimental plant was again placed in operation and ecological studies were initiated. As reported by Marphenin et al. (1997), there has been a gradual recolonization of the gulf and improvement of the heavily destroyed communities. Because of the mismanagement of this experimental tidal-electric station, its environmental impact is atypical.

15.4 ANNAPOLIS TIDAL-ELECTRIC PLANT

Conceived as a prototype project for the purpose of evaluating the operational characteristics of a large Straflo turbine and the potential of such turbines for large-scale tidal power developments in the Bay of Fundy and low-head, hydroelectric developments in Canada, construction of the Annapolis tidal-electric project was initiated in 1980 and the plant was commissioned in August 1984.

The project is located in the lower reaches of the Annapolis River at Annapolis Royal, Nova Scotia (refer to Fig. 15.13). It makes use of an existing causeway with a sluice-gate structure that had been constructed to replace old dikes as the means of protecting marshlands from the tides and to accommodate a highway access across the river. Before construction of the causeway, the average differences in water level in the river reservoir area was about 6.4 m but, after closure, the difference has amounted to only about 0.4 m. Thus, the presence of the structure has provided stable water levels immediately upstream. Although the upstream water levels are slightly higher with the tidal plant in operation, they do not exceed the previous maximum water levels during those periods when the river was in flood.

15.4.1 Existing Project Components

The existing causeway and sluiceway structures provided essential components for a tidal-electric development and, since the prototype testing of a large Straflo turbine required relatively large flows (of the order of 400 m^3/s) which are also available, the Annapolis site was judged to require the least cost for evaluating the operation and performance of the large diameter, Straflo turbogenerator (Baker, 1983). In this connection, the tidal setting also offered an advantage because of the continuously varying operating head, which would subject the unit to relatively severe operating conditions. On the other hand, the plant's operation became subject to the constraints of water control for marshland reclamation, which bears on the energy output and

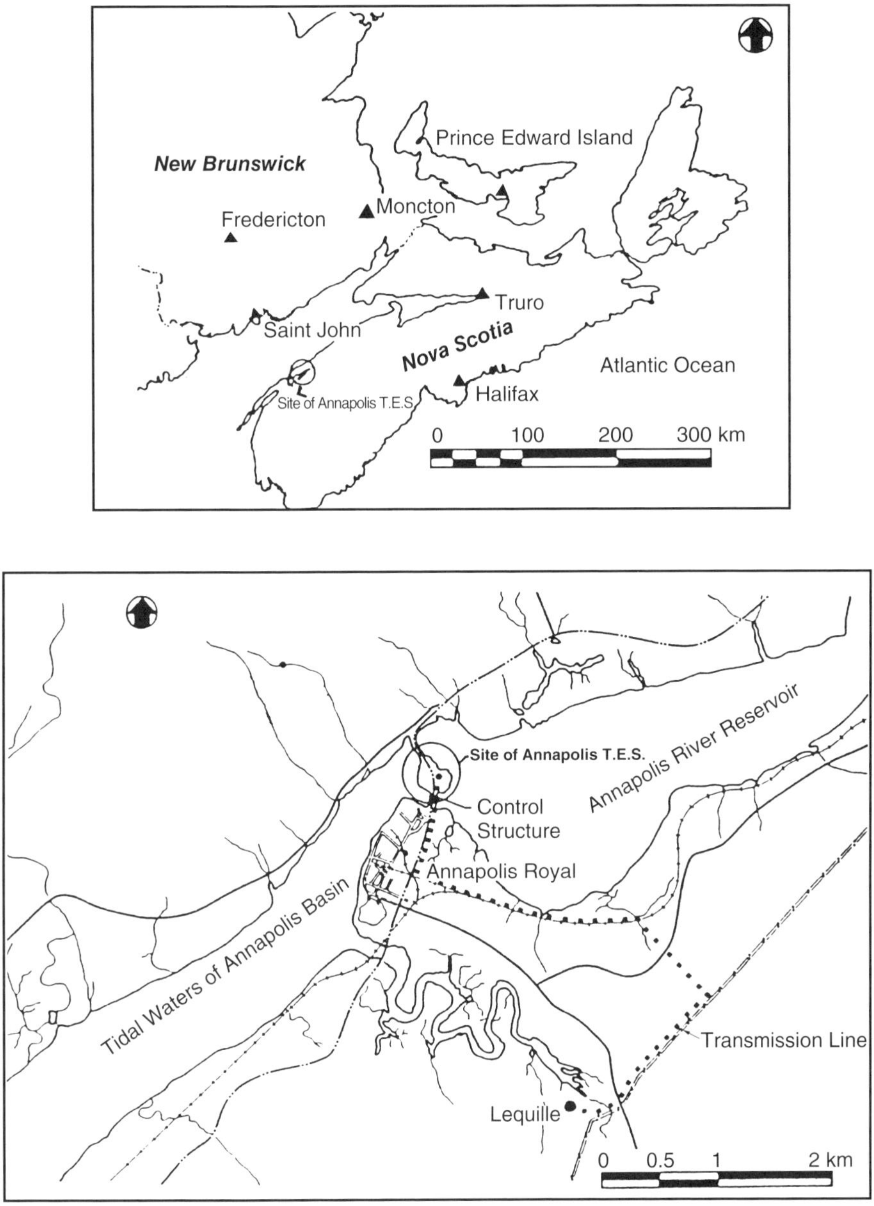

Figure 15.13 Location of Annapolis tidal-electric project.

project economics. Since the causeway was designed to limit the upstream water levels to 1.83 m maximum and 1 m average above geodetic datum, the plant is deprived of a portion of the heads available even from the modest tides at Annapolis.

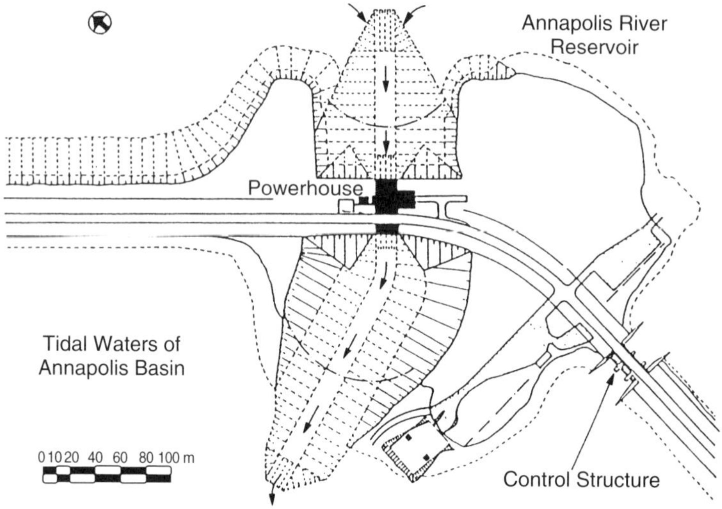

Figure 15.14 Site layout of the Annapolis Tidal Electric project.

The existing water control structure was of reinforced concrete and contained two, gated-sluice openings, each 9.2 m wide by 7.3 m high and one fish-pass opening, 3 m wide by 7.3 m high, at an elevation 1.83 m higher than the sills of the sluice openings. The water levels on the river side were regulated by electrically operated, vertical sluice gates. These gates and their seals were designed to prevent intrusion of seawater into the basin, and thus prevent flooding at high tide. The gates were raised to permit the discharge of river flow, as required, to maintain the water level in the reservoir. In order to be able to perform as an integral part of a tidal-electric development, the gates would have to be in the closed position during low tide and, therefore, the gate seals had to be modified to accommodate the new regime. The fish pass is not equipped with a gate so as to permit the movement of fish between sea and river, depending on tide levels. The maximum discharge at low tide amounts to approximately 500 m^3/s through the sluice openings and a maximum of 80 m^3/s through the fish passage. The plan view of the site is shown in Figure 15.14.

Due to the nature of the site, the powerhouse was constructed in the dry on Hog's Island (Fig. 15.15), which forms part of the causeway. This island had been created by the material from the control structure excavation. Figure 15.14 illustrates the site layout, and Fig. 15.16 shows the plant under construction. Material from the excavation for the powerhouse structure, the intake and tailrace were used to provide a seal on the upstream face of the causeway, which consisted of coarse rock-fill only. In its original condition, there was a greater flow through the causeway than through the existing fishway. Leakage and the pressure head at high tide caused some difficulties in fill placement, but this was surmounted by an appropriate sequence of placing fill material (Delory, 1986).

Figure 15.15 Aerial view of Annapolis Causeway showing Hog's Island and existing control structure. (Nova Scotia Power Corporation)

Figure 15.16 Annapolis tidal electric station under construction. (Nova Scotia Power Corporation)

15.4.2 Turbogenerator Features

The straflo turbine is desiged for single-effect operation, that is, designed to generate power during discharge from the reservoir into the sea. When the sea level is higher than the reservoir level, the water passages of the turbine are used as an orifice to sluice water into the reservoir during each tidal cycle. The powerhouse is 46.5 m long with a 15.5-m^2 intake opening transforming into a circle at the distributor. A similar transformation occurs in the draft tube, which measures 14.5 m × 11.1 m at the exit. Both intake and draft tube are divided by center piers and the intake pier supports the upstream combined thrust and guide bearing. The downstream bearing is a guide bearing only and is located in a housing supported by steel struts from the draft tube walls. A longitudinal section of the powerhouse and a cross-sectional view at the distributor gates are shown in Fig. 15.17 along with the main parameters of the Annapolis Straflo turbine.

The generator design is, to a large extent, governed by the turbine requirements (Baker, 1983). The generator poles are bolted to the face of the outer rim of the runner. There are 144 poles, as dictated by the rated speed of the turbine, supplied by static excitation fed through the upstream bearing and the runner's propeller blades. The stator is 13 m in overall diameter and supported by flanges resting on steel plates embedded in the concrete of the powerhouse walls. Rotational and transient loads are transferred to the powerhouse through two stops, one each at the top and bottom of the stator. One of the design problems, that of maintainence of a substantially uniform air gap over the working range of loads and temperatures, was solved by making the stator of as nearly as possible uniform stiffness throughout.

The insulation is designed to permit startup after immersion in saltwater, with only a preliminary freshwater flushing. The generator is air-cooled with excess heat transferred finally to the sea through air/water heat exchangers.

Stainless steel was used in critical areas only, that is, for blading, plating on wetted surfaces of hub and rotor rim, facing of sealing surfaces, and the contacting edges of guide vanes. All other structural components are carbon steel. An active cathodic protection system provides corrosion protection.

As discussed in Chapter 9, the straight-flow turbine uses the propeller blades as a spider to support the generator rotor. This requires sealing against water leakage out of the draft tube into the generator and into the main bearing housings (Douma and Stewart, 1981). The earlier straight-flow turbines used mechanical seals, and even though these units were considerably smaller in diameter, some problems were experienced. The rotor rim seals of the Annapolis Straflo turbine are of the hydrostatic type (Fig. 15.18). The seals consist of curved segments of special synthetic material pressed against polished stainless steel surfaces on the rotor rim. The seals are provided with filtered seawater for lubrication. Of the sealing water supplied, one-third flows radially outward past the seals and into a collector system surrounding the turbine, which then passes it over a measuring weir into a drainage sump. The

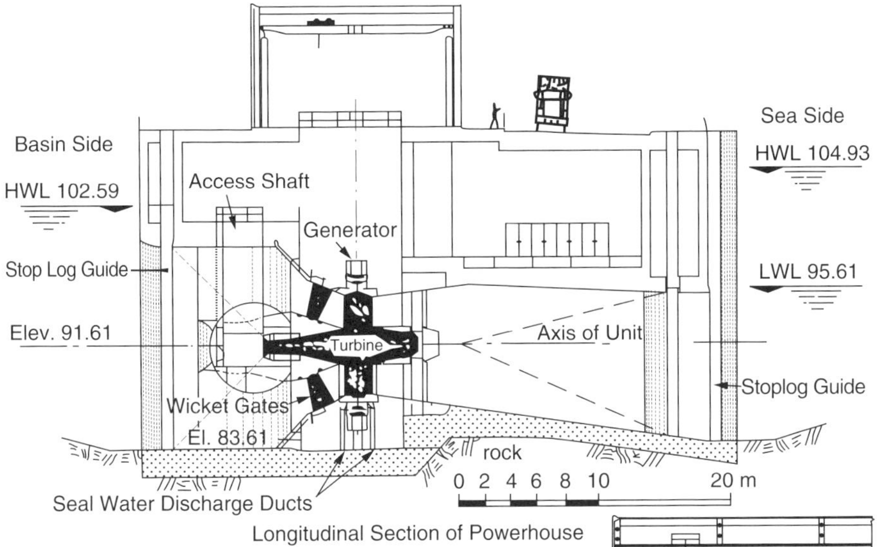

MAIN PARAMETERS OF ANNAPOLIS TURBINE	
Throat diameter	7.6 m
Speed	50 rpm
Runaway speed	98 rpm
Fixed propeller blades	4
Wicket gates	18
Rated head	5.5 m
Maximum output at rated head	19.6 MW
Maximum flow at rated head	407.5 m^3/s
Maximum output at 6-m head	19.6 MW
Maximum flow at 6-m head	383.6 m^3/s
Rim seals	Hydrostatic
Sealing water requirement	0.33 liters/m.s

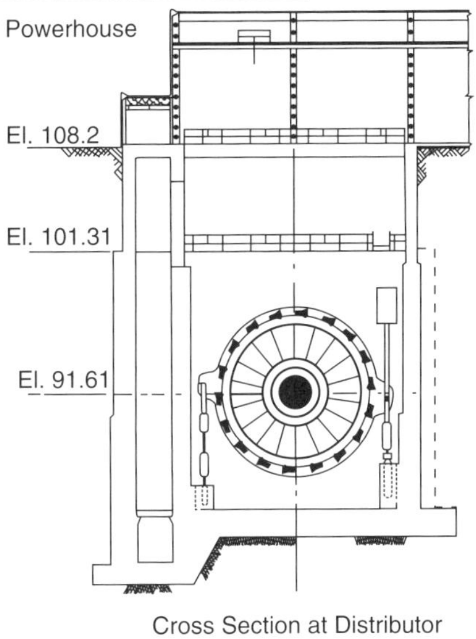

Figure 15.17 Sectional elevations of Annapolis powerhouse. (From Rice and Baker, 1987.)

other two-thirds is recirculated through the sealing water system. This arrangement provides an outward flow to prevent ingress of abrasive substances into the sealing system and also enables the monitoring of the rate of leakage water which would indicate whether or not there is any change in the seal condition. An additional tubular rubber seal is inflated to provide mechanical sealing when the turbine is at a standstill. The rotor rim seals are easily accessible and the seal segments can be inspected or replaced with a minimum of downtime (Baker, 1983).

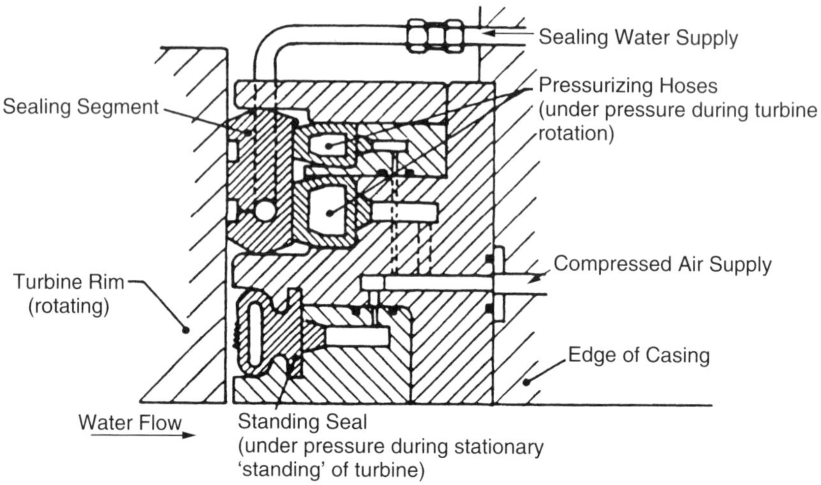

Figure 15.18 Detail of hydrostatic seal.

15.4.3 Operational Aspects

Since the station operates as a single-basin, single-effect tidal plant, the headpond is filled at high tide using both sluices and the turbine in its sluicing mode. When the tide has ebbed sufficiently to provide an optimal starting head, turbining commences. The operating sequence is illustrated in Fig. 15.19 for typical tidal conditions.

There is one pulse of generation per tide with a duration of 6 to 6.5 h. However, annual output is less than the design estimate of 50 GWh primarily because the level of the headpond, that is, the Annapolis River reservoir, is kept lower than its design level for environmental reasons, thus reducing the available head on the turbine.

Although cathodic protection proved completely adequate for wetted surfaces, it was found that, inside the powerhouse, salt air and frequently humid conditions necessitated a considerable maintenance effort to prevent surface corrosion of pumps, valves, pipes, and other equipment. This situation could be a problem in any tidal generating system located in a similar climate. Where possible, noncorroding materials and finishes should be used; for example, fiberglass pipes have proved satisfactory at Annapolis (Rice and Baker, 1992).

The tidal-electric plant is linked to the Nova Scotia transmission grid at Lequille (refer to Fig. 15.13) by means of a 69-kV line, approximately 5 km in length.

Figure 15.20 presents an aerial view of the Annapolis plant in operation.

15.4.4 Environmental Aspects

An increase in the rate of shoreline erosion occurred immediately following the increase in water level in the headpond and the river banks have now

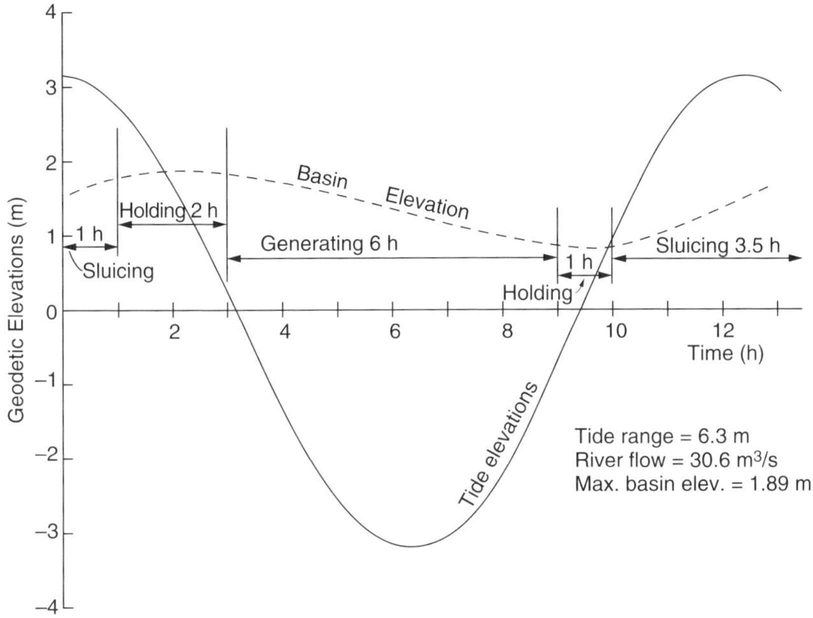

Figure 15.19 Typical operating cycle.

Figure 15.20 Annapolis tidal-electric station in operation. (Courtesy Nova Scotia Power Corporation.)

virtually stabilized to the higher water levels. Erosion rates have more or less returned to those prevailing before the development. In no case has erosion exceeded limits predicted by the engineering studies.

The many studies that had been carried out since the mid-1970s either related to fish habitat, populations, passage, and mortality in the Annapolis River delta, for the most part, with single issues in isolation or were directed to specific aspects of fish behavior. The varying methodologies applied and the sometimes conflicting conclusions made it difficult to make a reliable assessment of the station's impacts (Baker, 1992). An independent review commissioned by Nova Scotia Power in 1991, concluded that population studies over a period of three consecutive years of operation would be the most effective means of assessing overall turbine mortality rates since turbine effects, if significant, will ultimately manifest themselves in the population.

A 1995 survey of the shad population found that the individual fish size, age, and estimated number of spawning fish were somewhat lower than in 1985. This would be consistent with moderate turbine mortality, but it could also be part of the general malaise that has affected most species of commercial importance in the Atlantic fishery. The population is still large, and it would be reasonable to conclude that any effects due to the plant had stabilized by that time. The productivity of the headpond, which supports juvenile anadromous fish, has improved since the plant went into operation and exhibits more biological diversity. It would seem that the impacts on important fish species have, at worst, been only moderate and do not threaten the demise of fish stocks (Clark, 1997).

15.4.5 Summary

The few problems that have been encountered in the operation of the Annapolis tidal-electric station have involved auxiliary equipment rather than turbine design (Baker, 1992; Clark, 1997). None of the problems has appreciably affected the operation of the plant.

The peak efficiency curve associated with the fixed-blade, Straflo turbine would not be a disadvantage should dc transmission be contemplated, such as for the proposed Bay of Fundy projects, where asynchronous turbine operation would carry no economic penalty and, with variable speed, result in much the same characteristic as a double-regulated turbine. Moreover, the Straflo turbine is rugged, compact, and easily accessible, making minimum demands on powerhouse space and cost.

On the basis of the operating experience, the Straflo turbine would be a strong contender as prime mover in large tidal power developments, at least for those contemplated in the Bay of Fundy. The Annapolis unit has consistently achieved an availability of 99 percent, where *availability* is defined as the ratio of operating time to the total time when head equals or exceeds the minimum operating head.

CHAPTER 16

POTENTIAL DEVELOPMENTS

The events of the past several decades have clearly led to vastly altered perceptions of our energy futures. Energy has emerged as a critically demanding problem that, along with food and water, are the most significant, basic social problems challenging technological solution. Society is faced with the prospect that the continued, unrestrained burning of fossil fuels has initiated a change in global climate that will accelerate if this reckless approach to nonrenewable fuel resources continues. The *energy crisis* with which we have to deal may be one of supply for some countries but—for all countries—it is fundamentally a *crisis of adjustment.*

The major shift in attitudes regarding energy developments and the attractive features of tidal power provided a substantial impetus to the reappraisal of this renewable resource for use in large, modern, electrical utility systems. The development of electrical energy from the tides concerns the exploitation of an environmentally clean, renewable energy resource that is available in abundance, although there are relatively few locations in the world where the tidal range is large enough to make it economically exploitable under existing conditions. In addition to a sufficiently high tidal range, the site should also include a natural bay with an adequate area and volume and be so situated that the operation of the plant will not significantly reduce the tidal range.

The second half of the twentieth century has seen not only the development of three modern, small to medium-sized tidal electric stations, but also the investigation of various, and sometimes elaborate, schemes for the exploi-

Elements of Tidal-Electric Engineering. By Robert H. Clark

tation of tidal energy at some of the more obvious sites with large potential. The results of investigations undertaken since the 1950s, at some of the more interesting sites, are examined to highlight the variety of features that may be encountered and the particularly complex problems that may be associated with tidal-electric developments. Not all of the sites have been investigated to the same level of detail, and the current proposals for some sites are based on very meagre tidal data and other physical information.

Figure 16.1 shows the locations of those sites where tidal ranges are considered to be sufficiently large and that possess promising physical characteristics so that they have been given preliminary investigation or are being considered for study.

16.1 ARGENTINA

The southern coast of Argentina, from 36°S to 54°S, is influenced by some fairly large tidal ranges. The tides are semidiurnal, with an average range increasing in magnitude in a northerly direction from Tierra del Fuego to attain almost 10 m, then decreasing to about 4 m before increasing again to about 8 m in the vicinity of the Valdez Peninsula but decreasing again toward the estuary of the River Plate. The extraordinary size, proximity, and configuration of two large gulfs creating the Valdez Peninsula at about 43° south latitude, offers a possible and interesting site for a potentially economic tidal-electric development, even though the tidal range is not large—a maximum of about 8.5 m. Such a range would likely not justify the installation of a traditional single-basin development such as described by Fentzloff (1972) for the exploitation of the tidal range of the San José Gulf only, which has an area at high tide of about 780 km^2. A sketch of the region is given in Fig. 16.2.

The tidal flow in the vicinity of the Valdez Peninsula is absolutely unique, and the peninsula's configuration is such that there is a tidal difference of almost a half-period on either side of the 6- to 8-km-wide isthmus separating the two large gulfs of San José and Nuevo. The coastal morphology is apparently the cause of the peculiar propagation of the tidal current (Charlier, 1969). The Valdez Peninsula is a rather extensive projection into the Atlantic Ocean so that it requires about 4 h for the tide to round the peninsula; the tides are delayed a half period between the northern coast of the isthmus, that is, the San José Gulf and the southern coast, that is, Nuevo Gulf. A level difference of almost 16 m results, and this drop could be used to produce electricity dy dredging a canal through the isthmus and installing a double-effect tidal-electric plant with a probable installation between 600 and 1200 MW at the canal end in San José Gulf. This concept was given preliminary investigation by the Societé Grenobloise d'Etudes et d'Applications Hydrauliques (Sogreah) in 1959 and revealed no insurmountable construction difficulty. Apparently, the isthmus shows remarkable uniformity both structurally and petrographically. The sediments are predominantly Tertiary and constitute a compact and

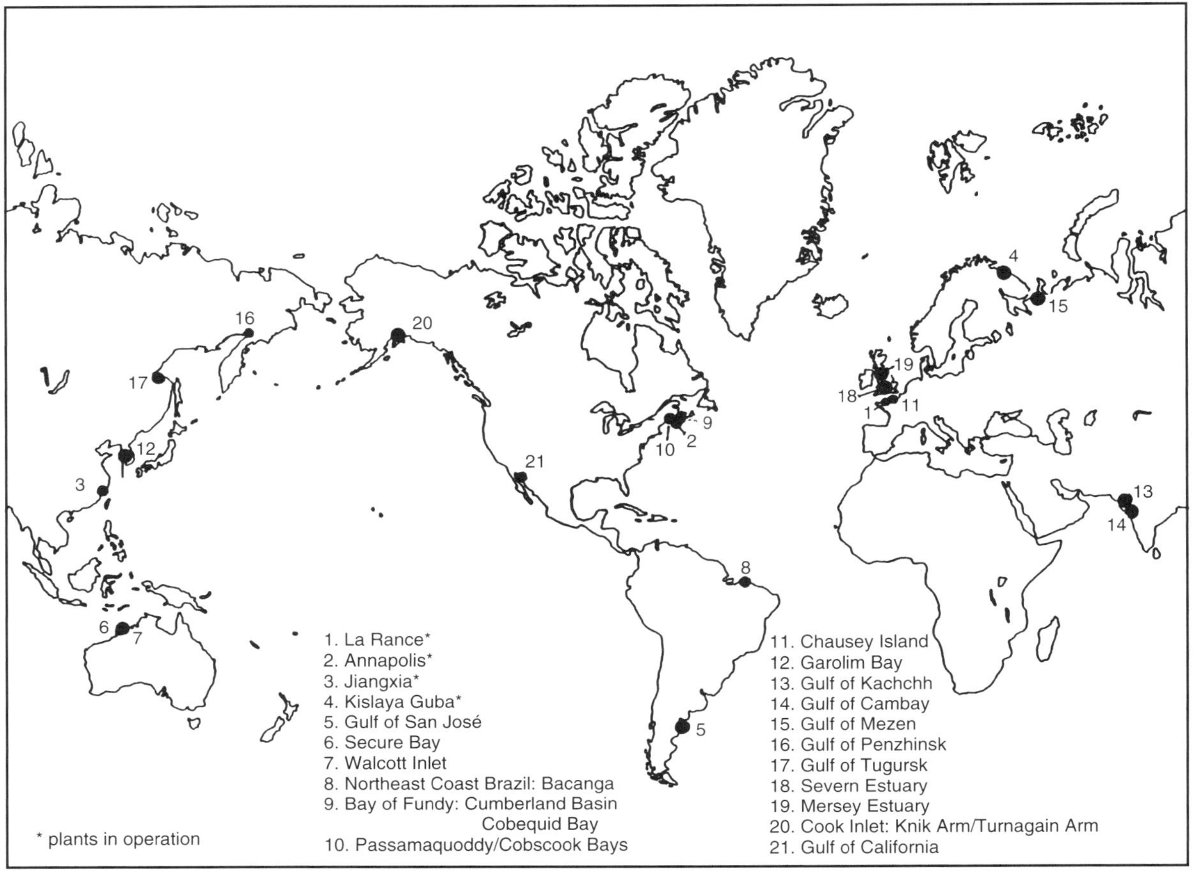

Figure 16.1 Location of major potential tidal power sites.

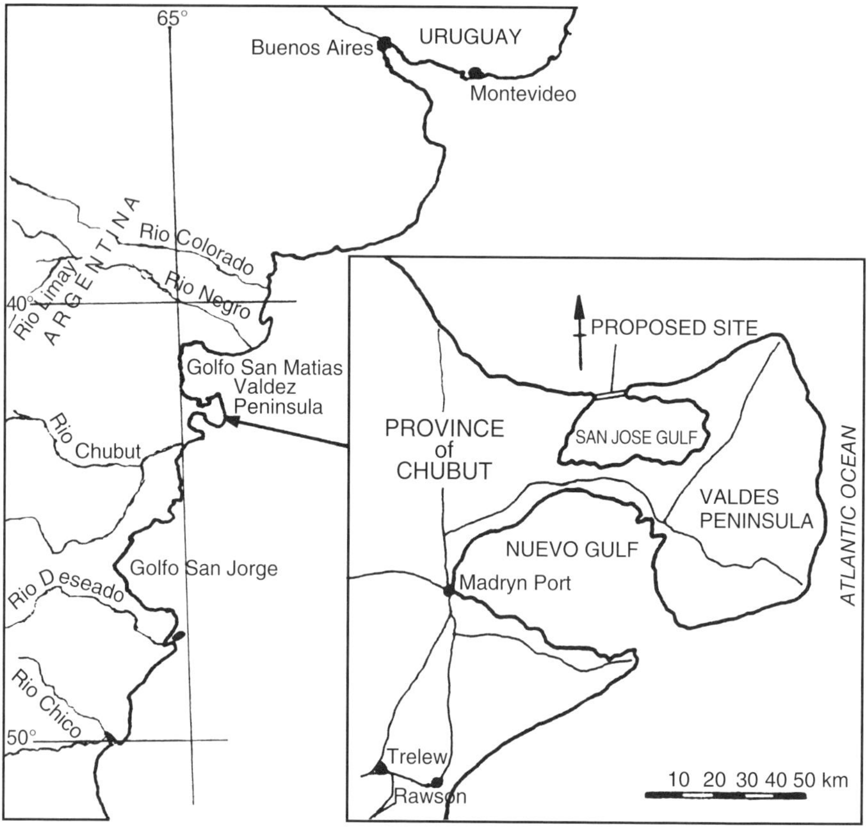

Figure 16.2 Valdez Peninsula and Gulf of San José.

impermeable strata, yet easily removed. The site is rather deserted and would be well-suited to the use of inexpensive methods to cut through the isthmus.

This unique tidal phenomenon has not yet been analyzed, and it could be possible that large-scale intervention could cause it to be greatly modified. Therefore, it would be absolutely necessary, prior to any major investigation of civil works, to understand the mechanics of the phenomenon. Currently, the available tide data are insufficient and inadequate for the necessary analyses. In this connection, measurements of currents and tidal amplitudes in the open sea will be necessary, at least to the limit of the continental shelf.

A review (Aisiks 1992) of a single-basin scheme exploiting the tidal energy potential in the San José Gulf, based on simplified operation studies, indicated that a near-optimal installation requires 180 generating units and 90 sluiceways incorporated in a 13.4-km barrage closing off the gulf. The powerhouse and sluiceway sections would consist of prefabricated concrete caissons and comprise about half the length of the barrage. Assuming a unit capacity of 28 MW with a runner diameter of 9.5 m, both the bulb-type and Straflo turbogenerators were considered. The installed capacity would be 5040 MW, and the mean annual energy generation would be about 9400 GWh. It was found that a development employing Straflo turbogenerators would be slightly less costly overall than one with bulb turbogenerators. However, the evaluation of new power plants in Argentina is now based on the short-term marginal value of generation, which, of course, favors the construction of projects requiring low, initial capital investment, such as conventional fossil-fired thermal units, so that a tidal-electric development would not likely achieve economic feasibility despite its obvious environmental advantages. Aisiks (1992) concluded that the tidal energy potential of Argentina would have a low development priority under the current economic conditions.

16.2 AUSTRALIA

The potential tidal power sites on the uninhabited Kimberley coast of the northern shores of Western Australia have been given cursory or preliminary study since the late 1950s. There are a number of possible sites that, if feasible, would have a large aggregate potential. Lewis (1963) reviewed the resources of the region and identified over 40 bays and estuaries in this rugged coastline (Fig. 16.3), where the mean spring tide ranges from about 5 to 11 m.

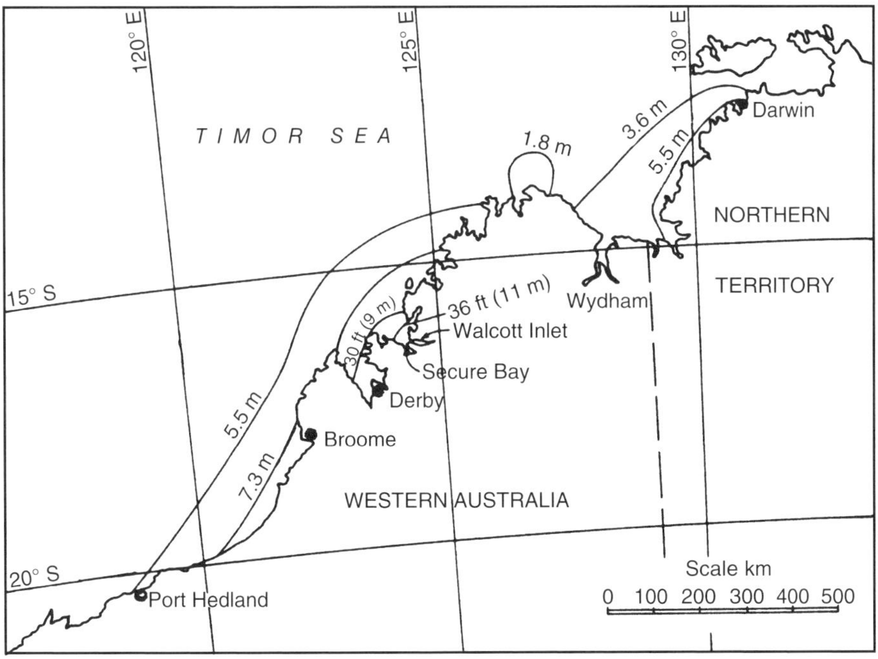

Figure 16.3 Kimberley coastline of Northwest Australia with tidal ranges.

Two adjacent estuaries, Walcott Inlet and Secure Bay (Fig. 16.4), with surface areas of 280 and 65 km^2, respectively, appear to have the most favorable configurations enhanced by their mean spring tidal ranges of up to 11 m. A 1965 report of a study, commissioned by the government of Western Australia, indicated that Secure Bay was the better of the two sites primarily because it offered a more appropriate scale for a first-stage development and a less severe tidal closure than Walcott Inlet. As determined by that study, the values of the parameters for development of these inlets as single-basin, ebb generation schemes are shown in Table 16.1 (Ackers, 1979). Construction of the main works was proposed to be carried out *in the dry* behind cofferdams.

Saunders (1974) reviewed the tidal potential of these sites and updated the estimated costs. He indicated that the construction time would have to be shortened by several years if a tidal-electric plant at Secure Bay were to be competitive with fossil-fueled generation under Australian conditions. This

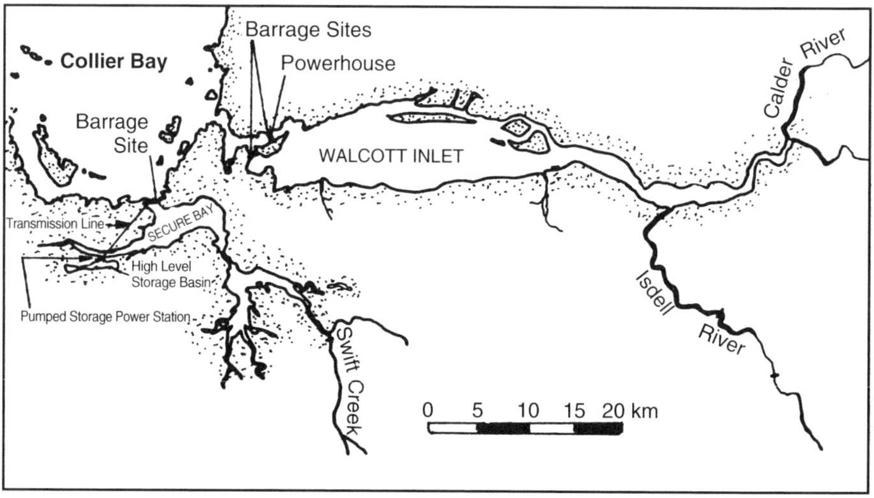

Figure 16.4 Single-basin development proposals for Walcott Inlet and Secure Bay. Note pumped-storage auxiliary facility for Secure Bay.

TABLE 16.1 1965 Development Characteristics at Secure Bay and at Walcott Inlet

Item	Secure Bay	Walcott Inlet
Installed capacity	570 MW	1,250 MW
Average annual energy output	1,650 GWh	3,940 GWh
Length of tidal barrage	1,850 m	2,500 m
Basin area at maximum tide	180 km^2	415 km^2
No. of bulb turbines (7.35 m dia.)	30	60
Sluice gate area	6,680 m^2	13,800 m^2
Excavation volume	4.3×10^6 m^3	58×10^6 m^3

shortening of construction time could be accomplished by the use of floated-in caissons with preinstalled equipment. It was also noted that a natural elevated basin adjacent to Secure Bay provides an attractive pumped-storage arrangement in conjunction with the single-basin arrangement (Fig. 16.4).

A further study was undertaken in 1976 (Ackers, 1979) with the objective limited to civil engineering aspects, and the main task to study the feasibility of closing the estuaries by new techniques that could speed the construction of the permanent works and reduce the costs of the 1965 proposals. The use of floated-in caissons was shown to be feasible, both hydraulically and structurally, and would considerably reduce the civil engineering costs; the use of *blind* caissons was also proposed to effect closure. Moreover, the study also confirmed the installation of 570 MW for Secure Bay, which could yield 170 MW of firm power utilizing the pumped-storage possibility.

Further investigations may well indicate the feasibility of developing a two-basin scheme using Walcott Inlet as the high-level basin and Secure Bay as the low-level basin. Such an arrangement is indicated in the sketch of Fig. 16.5. Because the area is very sprasely inhabited and there is no existing development in these estuaries, few problems of estuary management would be anticipated, but an environmental appraisal would have to be a component of any feasibility study.

In 1991, the Select Committee of "Energy and the Processing of Resources" presented its final report to the Legislative Assembly of Western Australia containing several recommendations relating to the potential future value of the tidal resources in the Kimberley region (Haws et al., 1992). It recommended, *inter alia*, that long-term plans should be made to harness Kimberley tidal power as one of the available renewable energy resources that Australia will need to employ in order to achieve sustainable development and to

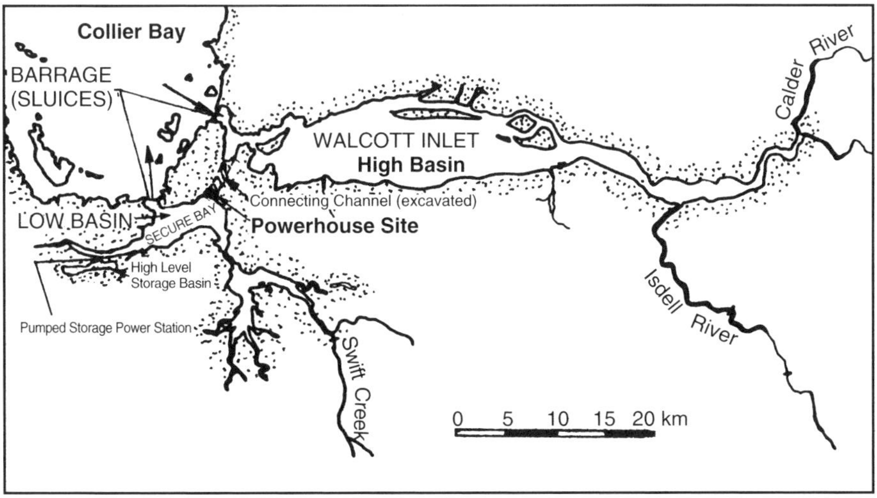

Figure 16.5 Linked-basin development proposed for Walcott Inlet and Secure Bay.

reduce greenhouse gas emissions, it also noted that any future natural gas pipelines from the northwest of Australia to markets in the south or east of the continent should be designed to allow the later transmission of hydrogen produced by electrolysis from tidal power in the Kimberley region. Proven gas resources in the region would likely meet current demand for some 200 years; other mineral resources are substantial and their extraction and processing could create a local demand for energy. The favorable long-term economics of tidal-electric projects could have a significant part to play in regional development and in husbanding the finite hydrocarbon resource. The hydrocarbons themselves would permit firming up the tidal energy with a gas turbine plant at low initial capital cost. Extensive prefeasibility investigations would be required in order to justify a costly, definitive feasibility study.

In 1994, the National Committee on Electric Energy issued a position paper that stated: "Tidal power will undoubtedly play a role in future power systems but significant development is unlikely to be economic for several decades and will need to deal with a range of environmental and heritage issues."

16.3 BRAZIL

In 1979, Centrais Eletricas Brasileiras SA (Eletrobras) commissioned a survey of the tidal power potential of the northeastern coast of Brazil (Fig. 16.6),

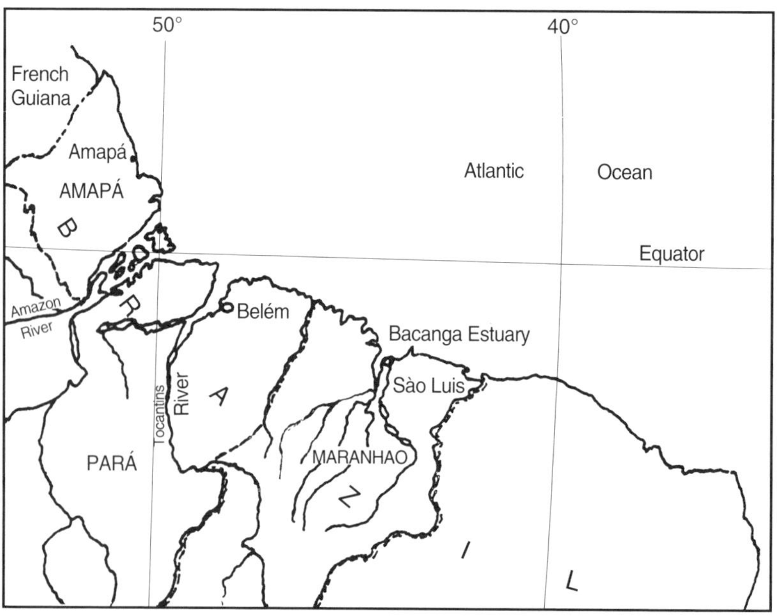

Figure 16.6 Northeastern coastline of Brazil.

where the mean tidal ranges vary from about 5 m along the coasts of Maranhao and Pará provinces to about 8 m along the coast of Amapá (near French Guiana) where a maximum of 13.5 m has been recorded in Maraca Harbour, near the town of Amapá (de Carvalho, 1982).

Forty-three sites were identified and a field reconnaissance carried out at each. Thirteen of the sites were selected for preliminary study. The total installed capacity at the 43 sites was estimated to be 27,000 MW with a potential annual energy production of 72 TWh, equivalent to a consumption of 432,000 bbl/day of petroleum. The unit cost of installation was estimated to vary between U.S.$1000 and $1500/kW (1980 price levels).

A design was prepared for a pilot project in the Bacanga estuary near Sào Luis, the capitol of Maranhao. The pilot plant would make use of an existing dam and spillways for an installation of between 5000 and 9000 kW in two-bulb turbogenerators capable of double-effect action, both generating and pumping, to provide an output of between 12 and 20 GWh, depending upon the installation.

16.4 CANADA

16.4.1 Bay of Fundy

Interest in exploiting the energy of the tides of the Bay of Fundy was expressed early in the twentieth century, but it wasn't until 1944 that the governments of Canada and the Province of New Brunswick sponsored an investigation into the exploitation of the tidal energy at the upper end of Shepody Bay (Department of Mines and Resources, Canada, 1945). That study indicated that a hydraulically linked, double-basin scheme, incorporating the estuaries of the Memramcook and Petitcodiac rivers (Fig. 16.7), could produce about 1.3 TWh annually but concluded the project was not economically justified. A number of investigations of limited scope were undertaken in the late 1950s and early 1960s in the Chignecto Bay and Minas Basin areas, but none of these was sufficiently definitive to establish the merits of development.

In 1966, the governments of Canada and of the provinces of New Brunswick and Nova Scotia jointly initiated a comprehensive investigation of the feasibility of large-scale tidal power developments in the Bay of Fundy. The details of the investigations and conclusions are set forth in the report of the Atlantic Tidal Power Programming Board (ATPPB) (1969). Once again it was found that, although technically feasible, it was not possible to exploit tidal energy at a cost competitive with energy from alternative sources nor would a Fundy tidal-electric development be economically feasible. However, the programming board recommended that further studies should be authorized when: (1) the interest rate on money drops sufficiently to suggest the possibility of an economic tidal power development; (2) a major breakthrough in construction costs or in the cost of generating equipment suggests the possibility of

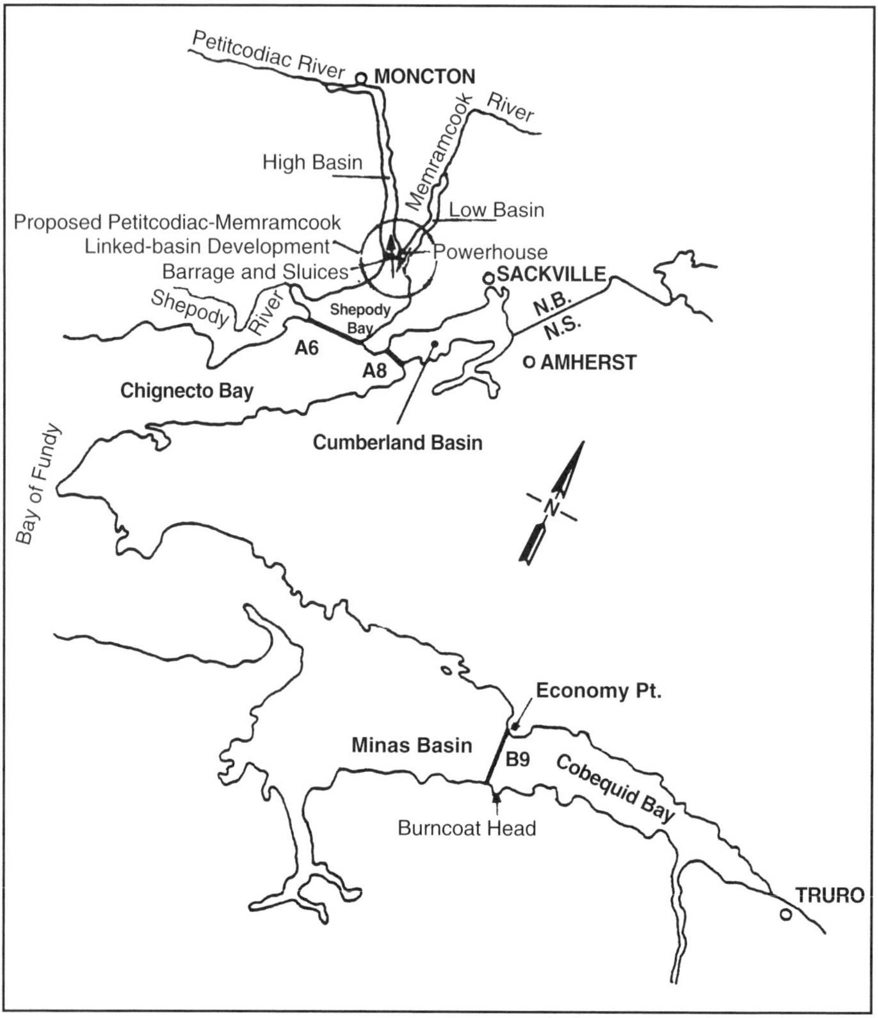

Figure 16.7 Location of selected tidal power sites for single-basin schemes in relation to the 1944 proposed Petitcodiac–Memramcook linked-basin development.

designing an economic tidal power development; (3) pollution abatement requirements magnify substantially the cost of using alternative sources of power; or (4) the alternative sources of more economic power supply become exhausted.

Subsequent to the submission of that report, not only did significant changes occur in the costs of alternative sources of energy but also a greater emphasis began to be placed on environmental considerations. Accordingly, the three governments established the Bay of Fundy Tidal Power Review Board in 1972 to undertake a critical review of the findings of the ATPPB in the light of the current and projected conditions and to recommend procedures for a reassessment of feasibility, if appropriate. The review board completed its preliminary reexamination in September 1974, concluding that the relative economic merits had converged sufficiently to warrant undertaking more comprehensive studies on a phased basis. In this way, the investigation could be terminated if the results of any phase indicated that tidal power could not become competitive with alternative energy sources. In submitting its conclusions to governments, the board pointed out that there were incentives that might encourage the market area to purchase tidal energy because reductions in fossil-fired energy production would result in: (1) reduced atmospheric and water pollution and (2) reductions in demand on decreasing oil and gas reserves that could be diverted to more critical areas. Although not specifically assessed in economic terms, the board also recognized that the transmission facilities associated with a tidal-electric development would undoubtedly offer additional opportunities for interchange of power within and between power systems of eastern Canada. Acting upon the recommendations of the review board, the governments of Canada and of the provinces of New Brunswick and Nova Scotia agreed in December 1975 that further investigation was desirable and in the public interest in view of the distinct possibility that tidal energy could be shown to be an economic resource in the Atlantic region, and authorized the appropriate investigations.

The agreement established a management committee to carry out the study program under the general direction of the review board. Most of the work was contracted with a number of eminent Canadian consulting firms. The results of the subsequent investigation are set out in the reports of the Bay of Fundy Tidal Power Review Board and Management Committee (1977). Material from those reports has been used extensively in the preparation of this text. Summaries of the investigation are also given by Clark (1978, 1979) and by Clark and Karas (1979).

The marketing approach used in the study is not unique to utility practice, that is, to determine how, and in what manner, tidal output could best be absorbed in the utility systems in which it would be operating. However, this differed from the approach that had almost been universally used in studies of tidal power up to that time—that of attempting to develop an "equivalence" between the output of a tidal power development and other conventional energy and firm power sources. The latter approach, applied to a single-basin scheme where the output occurs in the form of irregularly shaped blocks of energy, necessarily puts such a tidal energy source at a disadvantage. The approach adopted for the Fundy reassessment was based on systems considerations involving studies of alternative generation programs. At that time, it appeared likely that the utility systems in the Maritime Provinces would, in the future, have a large component of nuclear generation so that some storage retiming facilities would already be available in the system and, therefore, also available to a single-basin development operating under ebb tides.

The amount of tidal energy that can be absorbed by any system is related to that system's inherent retiming capabilities. This absorption capacity is a function of the system's size and load characteristics, the rate at which the output from other generating units can be backed off and displaced by tidal generation as well as the rate at which the units can be brought on line again when generation from the tidal plant ceases. Future system energy costs for various scenarios, including and excluding a tidal power development, were assessed. This utility system approach allowed for the absorption of raw tidal energy by the system and provided an implicit cost that, if not exceeded, could be applied by retiming of tidal energy either externally, for example, by pumped storage, or through direct, operational manipulation of the tidal power project.

The investigational program consisted of a series of technical and environmental assessments aimed at determining the competitiveness of tidal power. The work was focused on the optimization of selected sites to provide a consistent basis for the comparison of system costs of energy *with* and *without* tidal power.

An initial screening involved assessing the potential power and cost by parametric ranking of some 30 possible sites in the Canadian portion of the Bay of Fundy (Fig. 16.8). These sites embraced a range of capacities between 50 and 10,000 MW. System considerations indicated that the incremental size of generating units or additions anticipated by the utilities of an integrated

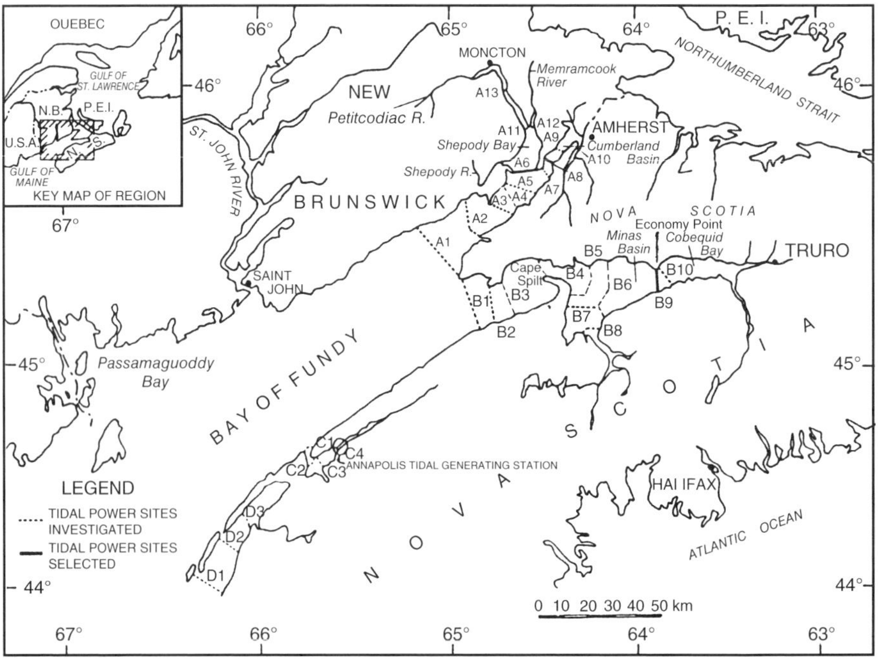

Figure 16.8 Location of Bay of Fundy tidal power sites investigated.

Maritime utility system would likely be about 4000 MW or larger by the time a tidal power development could be in service. Thus, potential sites that could offer considerably less than 1000 MW would likely not be of interest so that small projects were eliminated from consideration in the investigation; such small developments also ranked low on the economic scale.

As a result of the screening process, sites designated A8, A6, and B9 emerged as the best or preferred sites for single-basin schemes and could accommodate installations ranging from about 1000 to 4000 MW. The fact that these three sites were identical to those emerging from the screening process of the previous study (Atlantic Tidal Power Programming Board, 1969) confirmed the inherent superiority of those sites that were then studied in detail. Table 16.2 presents a summary of the characteristics for single-effect tidal power schemes at the preferred sites. Foundation conditions at these sites were found to be suitable for the use of "floated-in" caisson modules, which resulted in lower capital costs than conventional in situ construction behind temporary cofferdams.

The results showed that developments at sites A8 and B9 would be economic with a benefit–cost ratio of 1.2 and a breakeven period of between 30 and 35 years; a development at site A6 would be uneconomic. Although the review board recommended that funding be provided to complete detailed investigations and definitive designs, including detailed specifications, for a single-basin tidal power development at site A8 in Cumberland Basin, further action was deferred. Site A8 was recommended for a number of reasons, including the fact that the size of the project would tend to minimize any technical problems of introducing a new generation source into the Maritime region's electric supply and also that it would have no measureable effects on the tides beyond the limits of Canadian jurisdiction. The Cumberland Basin site (A8) is considered to be the preferred site for initial development not only based on socioeconomic considerations but also because: (1) it is the smallest of the three projects and would tend to minimize any technical problems of introducing a new generation source into the energy supply of the Maritime Provinces; (2) it is the largest and most economical site for which more than 90 percent of its output could be absorbed by the integrated system of the Mari-

TABLE 16.2 Characteristics of Selected Single-Effect Schemes

Item	Site B9	Site A6	Site A8
Total number of generating units	106	53	37
Total number of sluices	60	30	24
Number of spare generating units	6	3	2
Turbine diameter, m	7.5	7.5	7.5
Generator rated capacity, MW	38.0	31.0	31.0
Turbine rated head, m	7.5	6.5	6.5
Total installed capacity, MW	4,028	1,643	1,147
Annual output, MWh	12,653	4,533	3,423

time Provinces; (3) changes in tidal regime would not be measurable beyond the limits of Canadian jurisdiction; (4) regime effects would be less likely to give rise to significant adverse impacts than a development at B9; (5) capital investment requirements would be significantly smaller and would involve the least cost; and, although relatively small, the project is large enough to provide a substantial contribution to the energy requirements of the Maritime region of Canada. The plan and profile of the development proposed for site A8 is shown in Fig. 16.9.

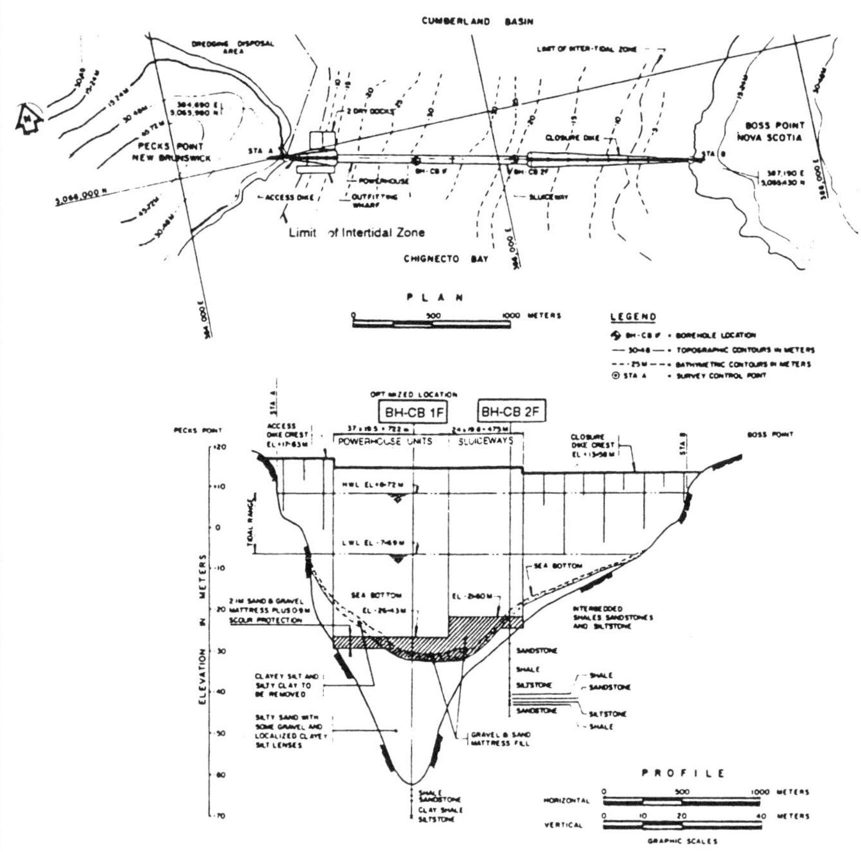

Notes
1. Elevations are referred to geodetic Survey of Canada datum.
2. H.W.L. refers to High High Water Large Tide plus 1.22m.
3. L.W.L. refers to Low Low Water Large Tide minus 0.76m.
4. Powerhouse – width of unit as shown
   – turbine diameter 7.5m
   Sluiceway – width of sluice as shown
   – sluiceway gate 12.2m x 12.2m
   Both powerhouse and sluices shall be constructed in modules of two units. One service bay shall be provided for every 16 powerhouse unites.
5. Installed capacity is number of units times 31 MW.
6. The service bay is incorporated with two generating units in a powerhouse module.
7. The length of dyke required for the dry dock (access dyke) shall be built by end-dumping. The balance of dyke section (closure dyke) shall be constructed in horizontal lifts from the sea-bed using marine techniques.

Figure 16.9 Single-effect layout for site A8, Cumberland Basin.

Subsequently, two separate updating studies were undertaken with the support of the governments of Canada and of Nova Scotia. The first was carried out by the Nova Scotia Tidal Power Corporation in 1981–1982 and the second by a private sector interest in 1984–1985. The results of the updating study by the Tidal Power Corporation (1982) are contained in its report, Update '82. The study concentrated on site B9, taking into account the changes that had taken place in electrical generating plant, fuel costs, rates of load growth, national energy policies, energy values, and other factors that could affect the results of the previous "reassessment." Revised designs and construction schedules introduced by the review would enable partial production by the plant about 2.5 years prior to commissioning, resulting in a significant reduction of the net cost. Under a reasonable marketing strategy, it was determined that the benefit–cost ratio for a development at B9 would be in the 2.4 to 3.0 range and that the ratio for a development at A8 would be about 2.0, applying a similar marketing strategy but based on the same design developed by the review board.

The 1984–1985 study, on the other hand, concentrated on site A8, revising the design and construction schedules. An overview of that study is presnted in a report by the Tidal Power Corporation (1985). Because the changes in design and in the relative cost of various components lead to changes in installed capacity, operating heads and turbine parameters, plant outputs were recalculated as an integral part of the design procedure. The approach to barrage construction developed for the 1982 update was applied. This involved placing sluice and turbine caissons in deep water, temporary sluice caissons or blind caissons in shallower water, and constructing dikes in water too shallow for caisson flotation. Horizontal closure would be effected by placing blind caissons, with currents minimized by maximum water passage through turbine, permanent, and temporary sluice caissons. Table 16.3 presents the updated design features at sites A8 and B9. The results for B9 were obtained by inference to those for A8. The updates of 1982 and 1985 reflected, in updated costs, the current knowledge of turbine/generator performance and improved construction techniques.

16.4.2 Ungava Bay

The tides in Ungava Bay, on the northern coast of eastern Canada, are of a magnitude to offer reasonably good development possibilities (Fig. 16.10). Although not as large as the tides in the upper reaches of the Bay of Fundy, which owe their magnitude to the narrowing and shelving of that bay, the mean tidal range increases to over 8 m in the southern reaches of Ungava Bay. Exploitation of this potential energy, which could be in the order of 50 TWh annually (Godin, 1974), would be integrated with that of the hydraulic energy of the five major rivers discharging a combined annual flow of some 4300 m^3/s into the bay from a plateau about 500 m above mean sea level. The potential hydroelectric energy of these rivers would undoubtedly have been exploited

TABLE 16.3 Updated Design Characteristics for Single-Effect Developments at Sites A8 and B9

Design Characteristic	Site A8 Cumberland Basin	Site B9 Cobequid Bay
Barrage length (m)	2560	8000
Dike length (m)	1340	1800
Number of powerhouse caissons	21	62
Number of deep sluice caissons	6	11
Number of shallow sluice caissons	4	35
Number of Straflo turbines	42	128
Diameter (m)	7.5	7.5
Rating (MW)	34	41.7
Installed capacity (MW)	1428	5338

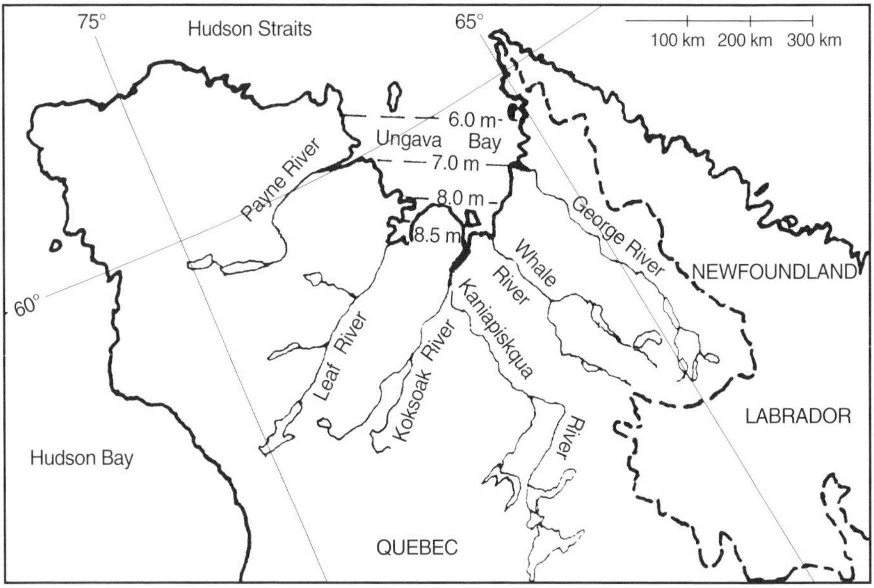

Figure 16.10 Ungava Bay and its hinterland. Broken lines denote mean tidal range in meters. (From Godin, 1974.)

many years ago had the region been closer to centers of population. Therefore, major barriers to even the consideration of development of these renewable sources of hydraulic and tidal energy are the transmission distances that would exceed 1500 km to major load centers as well as the lack of communications and development in this very sparsely populated region of Canada.

16.4.3 British Columbia Coast

The coastline of British Columbia has many inlets and bays that could be used as storage basins in tidal power developments. However, the mean tidal range along this coast is only 2 to 5 m, which is less than what would be considered favorable for exploiting the tidal energy. Nevertheless, in the late 1970s, a very preliminary study was carried out by the provincial hydro authority to identify potential sites along the coast. The major factors considered by this reconnaissance study for site selection were the magnitude of the tidal range, the size of the tidal basin for storage, and the size of the dam to close off the basin. Thirteen sites were identified (refer to Fig. 16.11 and Table 16.4).

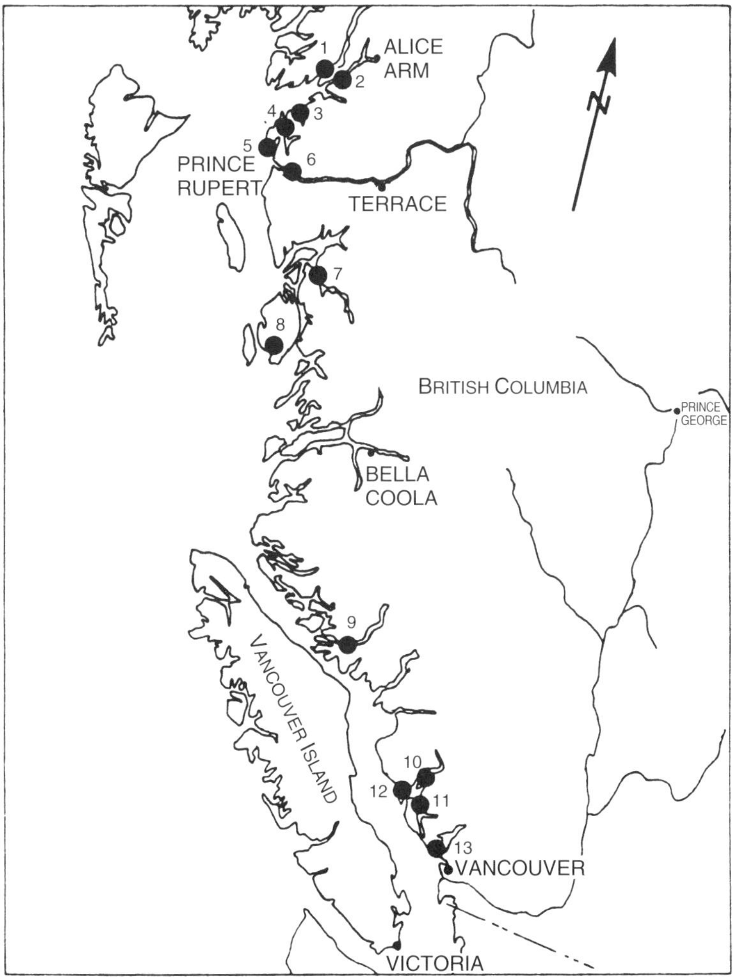

Figure 16.11 Coastline of British Columbia showing location of potential tidal power sites. (From British Columbia Hydro, 1979.)

TABLE 16.4 Potential Tidal Power Sites in British Columbia

Site	Mean Tidal Range (m)	Basin Area (km^2)	Installed Capacity (MW)	Annual Energy Output (MWh)
1. Portland Canal	5.0	209	738	2264
2. Observatory Inlet	4.7	212	662	2030
3. Khutzemateen Inlet	4.7	27	85	259
4. Work Channel	4.7	80	250	766
5. Prince Rupert	4.9	10	34	104
6. Skeena River	5.3	31	123	377
7. Gardner Canal	4.2	171	426	1307
8. Laredo Inlet	3.5	66	114	350
9. Knight Inlet	3.8	232	474	1452
10. Jervis Inlet	3.5	158	249	763
11. Sechelt Inlet	3.5	106	54	167
12. Jervis & Sechelt	3.5	430	677	2075
13. Howe Sound	3.3	282	395	1210

For the purpose of estimating costs of development, the two most attractive sites—Observatory Inlet and Sechelt Inlet—were selected. The former site is close to the border with Alaska and the latter is about 40 km northwest of Vancouver. The development of these two sites would provide about 2200 GWh of annual energy but, the unit cost of that energy would be about an order of magnitude greater than that from other generation options available to British Columbia.

16.5 CANADA–UNITED STATES

In the early 1920s, U.S. interests initiated a large-scale investigation of a two-basin scheme, involving the Canadian territory of Passamaquoddy Bay, to utilize the tides of that bay and those of Cobscook Bay, the latter being wholly within the United States. These bays are near the entrance to the Bay of Fundy. The tides range from 3.4 to 7.8 m, with an average tidal range of 5.5 m. The area contains a number of islands, thereby offering a number of alternate ways of constructing a two-pool system. The Canadian Biological Board reported adversely on this scheme in 1929 because of its effect on fisheries. As a result of continued interest in the exploitation of the tidal energy of these bays, both governments requested the International Joint Commission, established by the 1909 Boundary Waters Treaty between the two countries, to review all previous reports and to estimate the cost of carrying out a complete study in order to decide conclusively the engineering and economic feasibility of a large-scale international tidal power project utilizing Passamaquoddy and Cobscook Bays. Based on the board's report, submitted in 1950, the International Joint

Commission was given a reference in 1956 to "... determine the estimated cost of developing the international tidal power potential of Passamaquoddy Bay ... and to determine whether such cost would allow hydroelectric power to be produced at a price which is economically feasible." The study was also to consider the effect of the project on local and national economies, fisheries, and the like. The Commission's Engineering Board, which carried out the studies, reported on a hydraulically linked, two-basin scheme (Fig. 16.12)

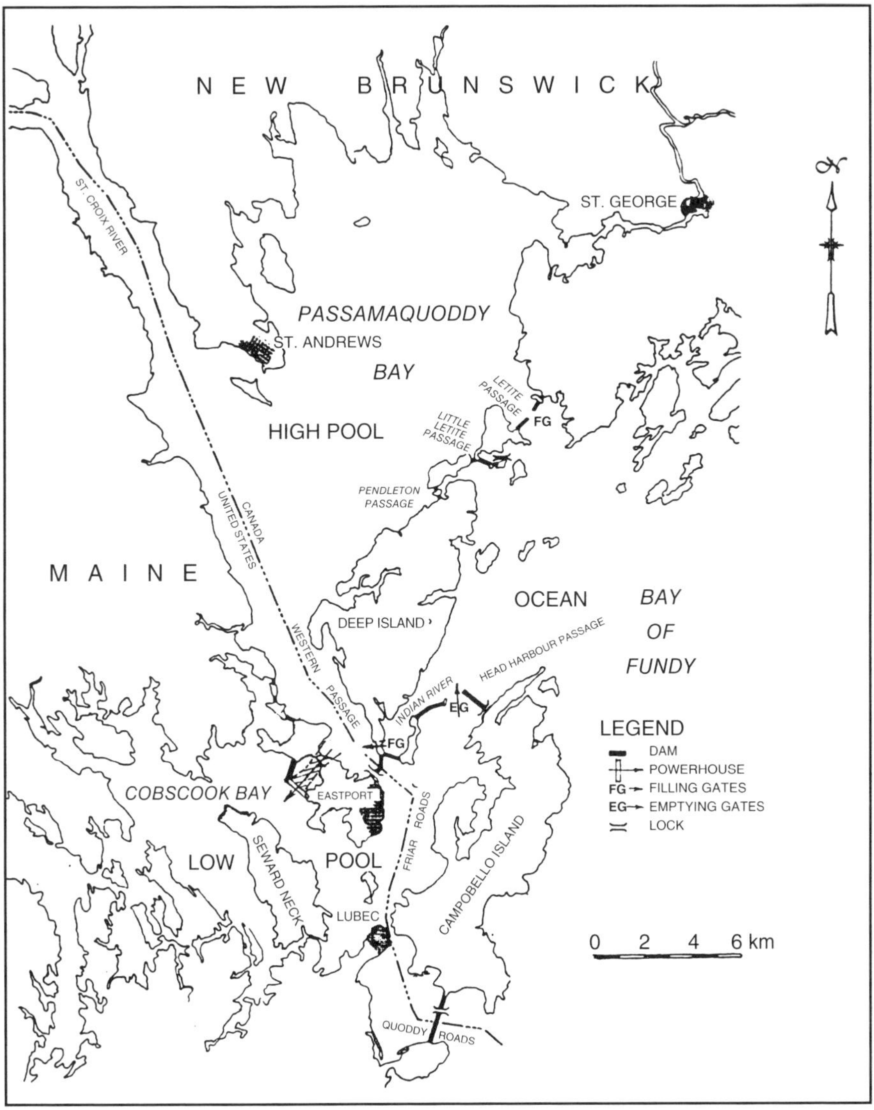

Figure 16.12 General arrangement of the proposed International Passamaquoddy Tidal Power Project.

with an installed capacity of 300 MW and a dependable capability of 95 MW (International Passamaquoddy Engineering Board, 1959). In its report to governments in 1961, the commission concluded, *inter alia*, that the tidal project, either alone or in combination with auxiliary power sources, would not permit power to be produced at a price that is competitive with the price of power from alternative sources. The benefit–cost ratio was substantially less than unity. The studies have been reviewed and updated periodically by the U.S. Corps of Engineers but the conclusion has remained the same.

16.6 CHINA

There are several tidal-electric plants in China but, with the exception of the Jiangxia tidal-electric plant described in Chapter 15, they are quite small, less than 1000 kW. There is little information on the tidal power potential along China's coast, and the following has been excerpted from Cheng (1985). A report on China's total exploitable tidal power potential was produced in 1981, based on an inventory of the potential of 156 bays and 33 estuaries with individual capacities of over 500 kW. The total exploitable tidal potential was assessed at 20,980 MW, corresponding to an annual energy production of 58 TWh. Of this exploitable tidal energy, 88 percent is located in Zhejiang and Fujian Provinces. Major potential sites for development include (refer to Fig. 15.8): Hangzhou Bay (2800 to 5500 MW), northern branch of Yangtze Estuary (500 to 800 MW), and Leqing Bay (550 MW). It was noted that the development of tidal power in China was still in the promotion and experimental stage and that none of the major sites is likely to be exploited in the near future.

16.7 FRANCE

Many schemes for exploiting the energy of the high tides of the Bay of Mont St. Michel have been proposed during the past five decades. With the commissioning of La Rance tidal-electric project in November 1966, however, official interest in further tidal power development in France waned, particularly in the light of the decline in fuel costs and the rise in interests rates at that time. However, the rise in fuel prices in the early 1970s sparked a renewed interest in tidal power as part of the quest for renewable energy sources. Several hydraulically linked, double-basin schemes along the Contentin Peninsula, embracing the Chausey Islands and Mont St. Michel came under consideration. A project in this area would, in addition to altering the basin's established ecosystem and changing the landscape, affect a very large population making a living from the sea (e.g., fishing and oyster and shellfish breeding) and from tourism. Appropriate studies were initiated to assess environmental impacts and effects on regional economic development.

The Chausey Island tidal project (Bonnefille, 1978) has many possible arrangements and sizes with estimated installed capacities ranging from 6000 to 15,000 MW, depending upon the characteristics of the proposed layout. The extensive layout shown in Fig. 16.13 was submitted by Albert Caquot, one of France's greatest engineers, in 1972. This proposal was probably the largest tidal power plant ever conceived up to that time—two basins of 1100 km^2, 35 km of dikes, and 50 km of floodgates—comprising two hundred, 30-MW units generating continuous electrical power during the working day and

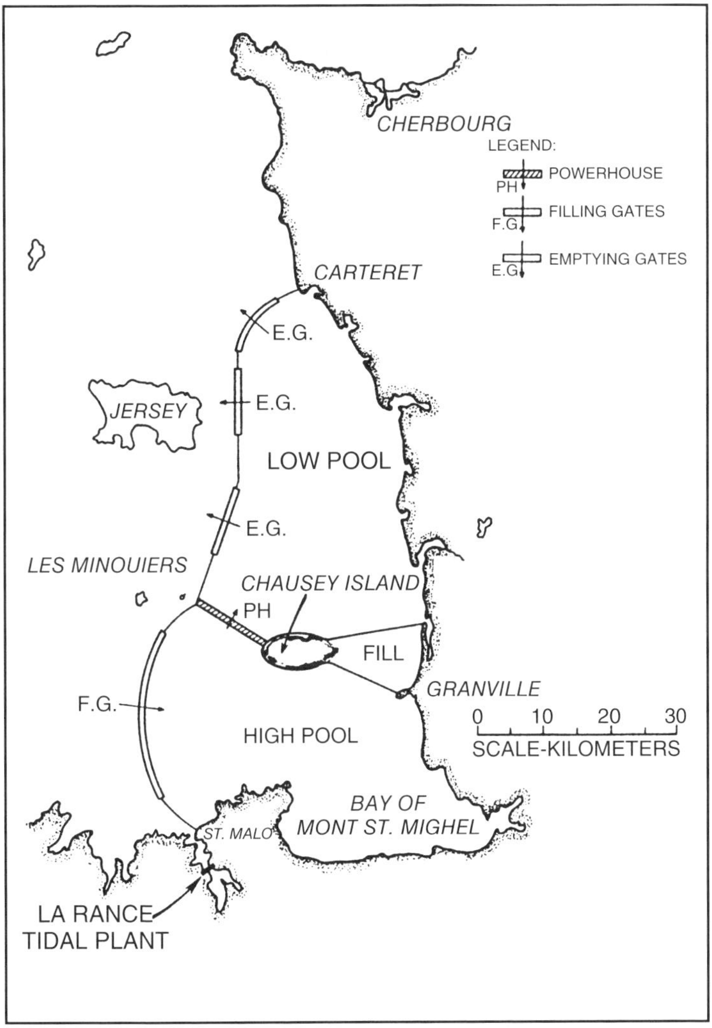

Figure 16.13 Proposed linked-basin scheme for Chausey Island site. (From Bonnefille, 1978.)

storing water by pumping at night, with an annual energy output of 34 TWh and with an assured output of 3900 MW.

16.8 INDIA

The development of tidal power in India has received sporadic interest during the past few decades. Development possibilities have so far been identified at three locations: the Gulf of Kachchh and Gulf of Khambhat on the west coast in Gujarat Province and in Sunderbans along the east coast in West Bengal. According to the tidal characteristics shown in Table 16.5, the average ranges on the west coast are not very large; however, the average ranges on the east coast are substantially smaller.

An assessment of development possibilities carried out under the sponsorship of the United Nations (1975) proposed single-basin schemes to develop the energy potential of the two estuaries in Gujarat Province (Subrahmanyan, 1978; Wilson, 1979). A single-basin installation in the Gulf of Khambhat might range from about 5000 to 7000 MW while that to exploit the tidal energy of the Gulf of Kachchh could be up to 1200 MW. A linked-basin scheme was also proposed for the latter estuary with an installation of 600 MW, but, based on preliminary estimates, the cost of energy generation by the two-basin alternative would be almost double that from the single-basin scheme, although it would provide some firm capacity and energy. Owing to the smaller tidal ranges and possible basin areas, development possibilities in the Sunderbans region would likely be small scale, on the order of 2 MW, although one site there might support a 15-MW installation.

The problem of integrating the intermittent output of a large, single-basin scheme, such as that suggested for the Gulf of Khambhat (Fig. 16.14), with the electrical power system of Gujarat Province, which is predominantly thermal based, would likely pose an insurmountable problem for some time (Subrahmanyan, 1978). In addition, other factors, such as the large amount of silt carried by the major river tributary to the gulf, will require detailed investigation. Very extensive civil works would also be required.

TABLE 16.5 Tidal Characteristics in Gulfs of Khambhat and Kachchh

Location	Spring Range (m)	Neap range (m)	Average range (m)
Bhavnagar (Gulf of Khambhat)	10.29	3.6	7.0
Navlakhi (Gulf of Kachchh)	7.17	3.74	5.45
Kandla (Gulf of Kachchh)	6.32	3.68	5.00

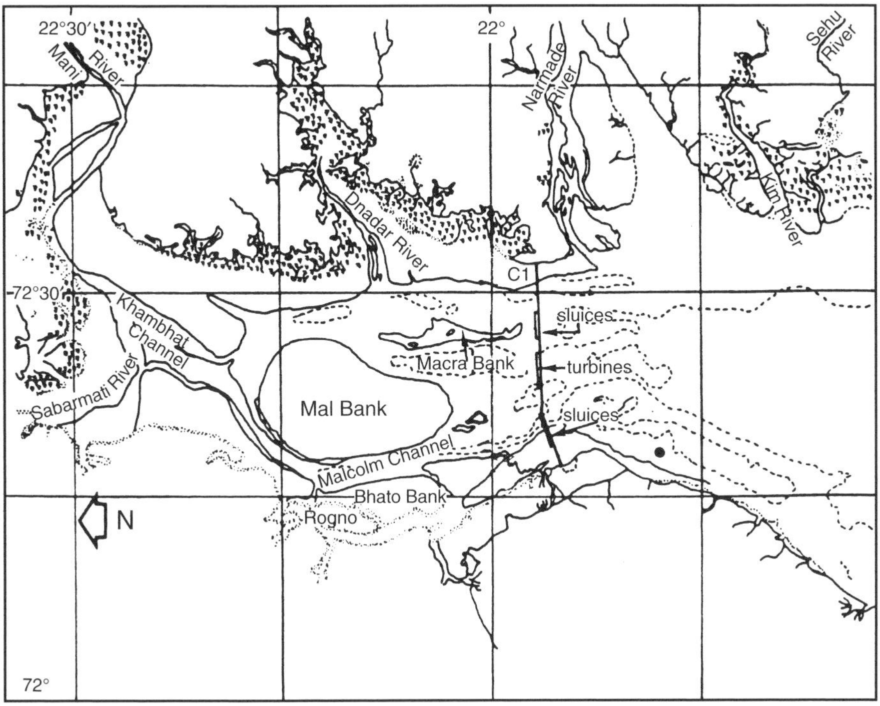

Figure 16.14 Proposed single-basin scheme—Gulf of Khambhat.

However, there is a current investigation underway involving a multipurpose development, namely exploiting the tidal energy to raise the freshwater from the rivers tributary to the gulf to an elevation about 60 m above mean sea level in order to provide irrigation water for agricultural lands in the Khambhat region. A barrage across the mouth of the gulf is proposed to store the water from the major rivers, including the Narmade. The tidal basin, to which the inflow of the tidal waters would be confined, would be completely segregated by dikes from the inflowing freshwater to the reservoir created by the barrage across the mouth of the gulf. The size of the basin isolated by the dikes and the allowable drawdown within the tidal basin would determine the capacity of the installation. Thus, the freshwater would be impounded by the barrage and would then be pumped periodically to a higher reservoir for irrigation by gravity of the agricultural lands in the region.

A feasibility study to exploit the energy potential of the Gulf of Kachchh was launched in 1982 and completed in 1987. The development proposed for this gulf would be a single basin with the main tidal barrier, about 3.25 km long, constructed across Hansthal Creek, using floated-in caissons to form the powerhouse, sluices, and navigation locks (Fig. 16.15). Three subsidiary barriers of about 280 and 300 m in length would be constructed across the Sanu,

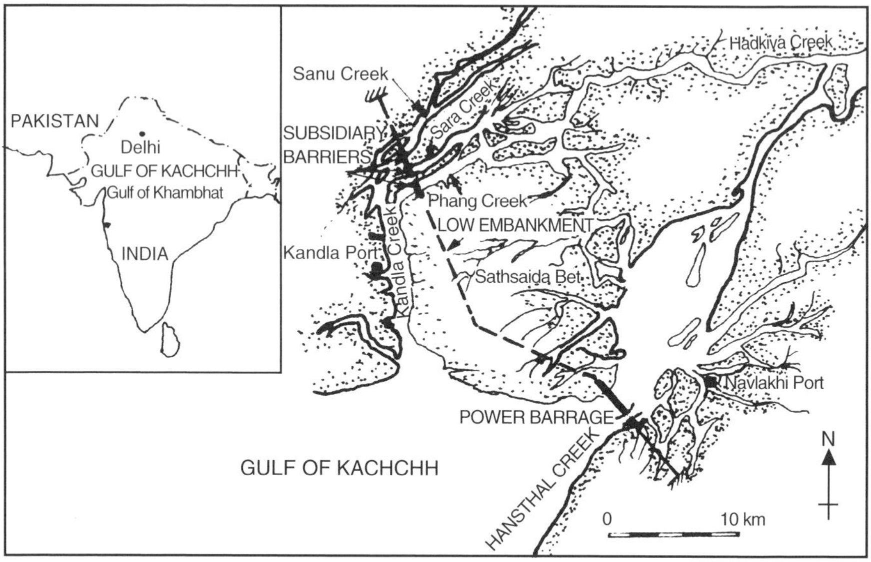

Figure 16.15 Proposed single-basin scheme—Gulf of Kachchh.

Sara, and Phang creeks, respectively. In addition, a low embankment about 55 km long and about 5.5 m in height would have to be constructed across Sathsaida Bet to isolate the tidal basin from the main gulf since the Sathsaida Bet becomes submerged during high tides. The barrage would be connected to high ground at either end so as to provide a permanent roadway across the gulf. The installation would consist of 36 adjustable-blade bulb turbines, 8.5 m in diameter and with a capacity of 25 MW under a rated head of 5 m; the maximum head would be about 7.5 m. The total installation of 900 MW would yield an annual output of about 1690 GWh (Sharma, 1986, 1989).

16.9 KOREA

The exploitation of the tidal energy in the many bays and inlets along the west coast of Korea, notably Kyonggi, Asan, Garolim, and Cheonsu bays (Fig. 16.16), has been the subject of sporadic investigation for several decades. In these estuaries, which open onto the Yellow Sea, the mean tidal ranges are about 5.7, 6.1, 4.8, and 4.5 m, respectively. The Korea Electric Company commissioned the first phase of a long-term development plan in 1978 aimed at identifying the tidal power potential of the west coast and the selection of the best site. The study identified 10 possible sites and recommended Garolim Bay for a feasibility study.

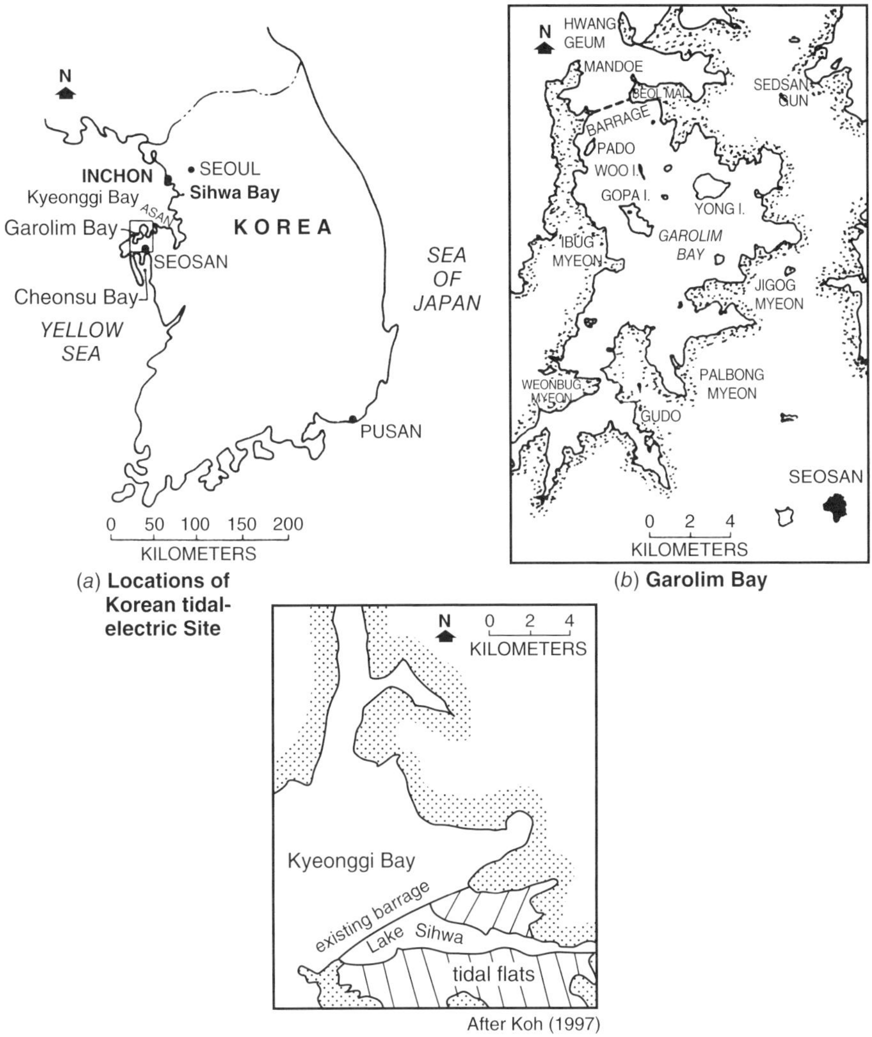

(*a*) **Locations of Korean tidal-electric Site**

(*b*) **Garolim Bay**

(*c*) **Sihwa TPP Location**

Figure 16.16 Tidal power sites in Korea.

16.9.1 Garolim Bay

Garolim Bay is situated in the southern part of Kyonggi Bay, as shown in Fig. 16.16*b*. It has an area of about 50 km^2 at mean sea level; the mouth of the bay is about 2 km wide with a maximum depth of 25 m. The tides of Garolim Bay are of the semidiurnal type with a large diurnal inequality; the mean tidal range is about 4.7 m, with a mean spring range of 6.6 m. A feasibility study was carried out in 1981 (Sogreah, 1981) to establish a design adapted to the physical conditions prevailing at the site and the technical and economic characteristics of the Korean electric system. The barrage would be constructed at the site shown in Fig. 16.16 using floated-in concrete caissons containing turbines and sluices, embankment dikes, a navigation lock, and a switchyard. The turbines would be located on the western side of the barrage in the deep-water channel. A reassessment and update of the proposed development was undertaken in 1986 (KORDI, 1986) incorporating the results of further oceanographic field work that had been carried out during the interval. A number of technical refinements and economies were introduced and the generating installation, reoptimized. The results of the reassessment indicated an installation of twenty, 20-MW double-regulated, bulb turbines, 8 m in diameter, operating in the single-effect mode and with no pumping. The installation of 400 MW would provide an annual output of about 836 GWh. Although the reassessment resulted in about a one-third reduction in construction cost, the economic feasibility could not be justified under the stipulated discount rate of 10 percent (Song, 1992). A Garolim tidal-electric plant could provide a major source of renewable energy, providing a useful diversification for the Korean electric generating system. It was noted that a development could incorporate other purposes, such as recreation facilities, marine aquaculture, and a small harbor and dock marina inside the basin that would enhance its benefit–cost ratio.

16.9.2 Sihwa Bay

The high tidal ranges associated with the gentle sea-bottom slopes create broad tidal flats along the Korean west coast that have long been exposed to an industrial use for the achievement of economic growth, particularly during the last half of the twentieth century. For example, Sihwa Bay became an artificial lake after a 12.6-m seawall/barrier was constructed at the outlet of Sihwas Bay and completed in 1994 (refer to Fig. 16.16c). The land/tidal flats made available by the structure amounted to about 170 km^2. About one-third of this land is now used for industrial purposes and the rest for agriculture. However, the pollution problem in this artificial lake, formerly a tidal channel, has raised much concern. To rectify this problem, and based upon the result of a feasibility investigation, the government of South Korea decided, in 2005, to exploit the energy of the 5.7-m tides of the former Sihwa Bay. The Sihwa Tidal Power Plant, which will be integrated with the existing structure, will consist of ten, bulb-type turbines with direct-driven generators. The output of each unit will be 26 MW and the plant is to be operated in single-effect mode, that is, flood generation (see Chapter 1). The estimated cost of the project has been stated as US$250 million and is scheduled to be completed by 2009. Not only will the project develop electrical energy but the existing poor water quality of Sihwa Lake will be significantly improved by the exchanges of water with the sea through the tidal power barrage (Koh, 1997).

16.10 MEXICO

16.10.1 Gulf of California

Although there is a spring tidal range of 7m at the estuary of the Colorado River at the northern end of the Gulf of California (Fig. 16.17), the site is in an area of extreme seismic activity. A tidal barrage would lie astride the San Jacinto Fault in the trough between the Pacific Plate and the North American Continental Plate. Not only is the trough partly filled with Tertiary age silts, sands, and gravels and thick deposits of Quaternary alluvium and lake deposits, but the bedrock is badly faulted, fractured, and tilted (Wayne, 1977).

Measures for the economic exploitation of the energy potential were investigated by Mexican engineers. However, in view of the foregoing geological problems, it was not proposed to construct a barrage across the Colorado estuary but rather to form an artificial embayment on (a laqoon) the coast of Sonora State, that is, the east side of the Gulf of California. Initial studies indicated an energy potential of 3.8TWh.

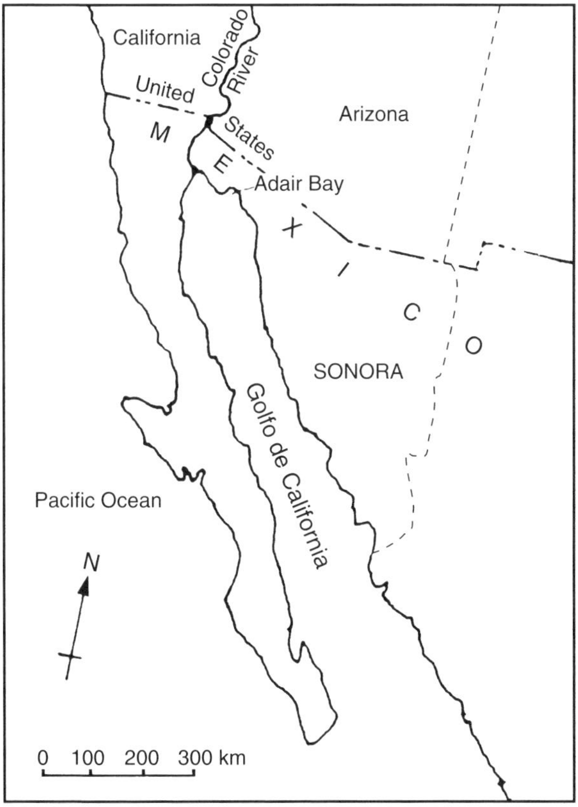

Figure 16.17 Gulf of California.

16.11 RUSSIAN FEDERATION

The tidal power potential on the northern coasts of Russia has been under active consideration for a number of decades. There are several locations where the average tidal range is fairly high and the coastal configurations conducive to the exploitation of the energy. The basic objective of constructing the Kislaya Guba tidal-electric station was the more cost-effective development of the large potential sites on the coasts of the White, Barents, and Okhotsk Seas (refer to Figs. 16.18 and 16.19).

The following information has been extracted from reports of L. B. Bernstein (1974, 1979, 1986, 1992). Since about 1938 to the present, all tidal power surveys, investigations, and designs to harness the potential along Russia's northern coasts have been carried out under the direction of Dr. Bernstein, from the Hydroproject Association, Moscow. The most promising large sites that have been identified are listed in Table 16.6 (Bernstein, 1992). Detailed information is consolidated in the 1996 English translation, edited by E. Wilson and W. O. Song, of Bernstein's *Tidal Power Plants*.

The results of mathematical modeling of the effects of operation of tidal-electric plants at the latter three sites are discussed in a study by Nekrasov and Romanenkov (1997). In general, it was concluded that the effects are proportional to the size of the basin isolated by the barrage.

16.11.1 Lumbovsk Bay

This bay is situated on the Kola Peninsula on the Barents Sea coast (refer to Fig. 16.18). It has an oval shape with a water surface area varying from 56 to 70 km^2. A development could have a total capacity of 670 MW with an installation of 64 turbogenerators, 10 m in diameter. The annual energy output would be about 2 TWh. However, a development could not be economically justified primarily because of the fairly low or modest tidal range.

16.11.2 Mezen Bay

This bay is located near the entrance to the White Sea (Fig. 16.18) and provides the main resource of exploitable tidal power on the European coastline of Russia (Bernstein, 1992). A possible scheme would cut off an area of 2640 km^2 in the Mezen Estuary by means of an 85.6-km-long barrage as shown in the figure. The barrage would be in the form of a seaward trapezoidal projection that will provide room for the 19-km-long powerhouse at depths requiring no excavation. The installed capacity of this site would be 15.2 GW with an annual energy output of 15 TWh from 800 double-regulated bulb turbines of 19 MW capacity, 10 m in diameter. The powerhouse would be formed by 200 caissons, each containing four units. The barrage would also include caissons consisting of two fish-passing facilities and two locks. The remainder of the dam (66.3 km) would be of solid rock-earthfill construction with a sandy

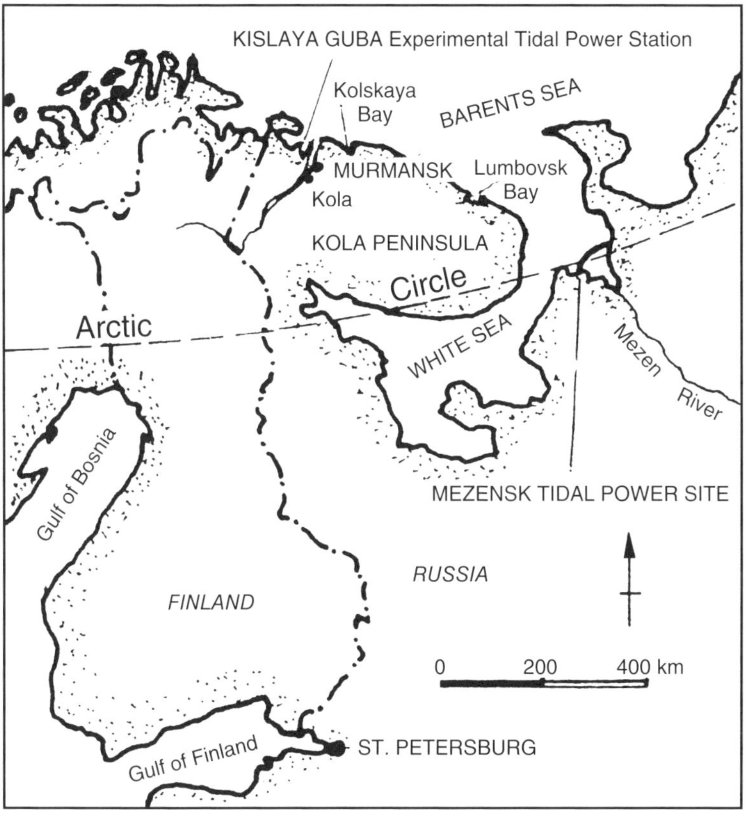

Figure 16.18 Location of proposed Barents Sea and White Sea tidal power projects.

core; the slopes would be stabilized with reinforced concrete slabs floated to position. The enormous volume of work to be carried out in the construction of a Mezen tidal-electric station will require considerable study and innovation to reduce costs.

16.11.3 Okhotsk Sea

The feasibility of tidal-electric plants in Tugur Bay and the Gulf of Penzhina has been a subject of investigation during the past two decades. High spring tides of 13.5 m in the northern region of the latter gulf are comparable to those in the Severn Estuary and the Bay of Fundy. One possible scheme investigated to exploit this energy envisions a barrage at the narrowing of the gulf (refer to Fig. 16.19), which could accommodate 568 turbines of 10 m diameter and 19.8 MW capacity over a section 32.2 km long and 26 m deep with 920 turbines of 7.5 m diameter and 11 MW capacity placed in the section of lesser depths. The powerhouse would be formed by four-unit, two-tier caissons whose cross

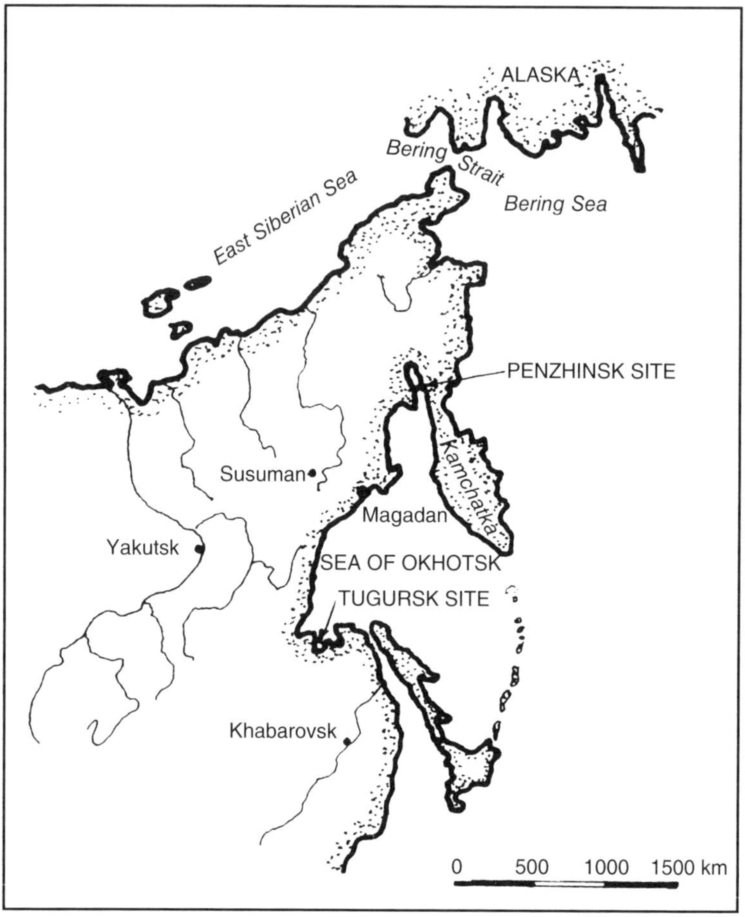

Figure 16.19 Location of tidal power sites—Sea of Okhotsk.

TABLE 16.6 Potential Tidal Power Sites—Russia

Site	Mean Spring Tidal Range (m)	Area of Basin (km^2)	Installed Capacity (GW)	Annual Energy Yield (TWh)
Lumbovsk	4.20	70	0.67	2.0
Mezensk	5.66	2640	21.00	15.0
Penzhinsk	6.20	6788	21.4	71.4
Tugursk	7.25	1120	6.80	16.2

section is illustrated in Fig. 16.20. The total capacity of the scheme would be 21.4 GW with an annual output of about 71.4 TWh.

The region of the Gulf of Penzhina is subject to severe climate conditions with an average annual temperature of –6.5°C and minimum temperatures of –50°C. Such weather contributes to the icing problems and the gulf becomes an ice-making factory. Beginning in November, drifting floes cover the whole of the water area and, under the action of the strong tidal currents, reaching 2 m/s, and the water level fluctuations, the floes are continuously broken and reefed. The height of the hummocks on floating ice may reach 3 to 4 m. It is considered that the exploitation of this huge volume of tidal energy from the Gulf of Penzhina may be of interest in the more distant future.

The natural conditions of Tugur Bay, on the south coast of the Okhotsk Sea, are relatively favorable for the construction of a tidal-electric plant. The climate, slightly milder than that in the northern regions of the Gulf of Penzhina, exhibits an average annual temperature of –3.6°C with variations from +20°C to –25°C. The bay is protected against heavy ice floes by the ridge of islands at its entrance. Since the physiographical and geological conditions would be favorable to the development of a tidal power plant, the government decided in 1987 to prepare a technical and economic appraisal of a Tugur tidal-electric plant.

The site selected is at a narrowing of the bay where the tidal range is larger than at the mouth of the bay and the barrage size would be smaller. Based on an extensive design analysis and optimization procedure, double-effect operation was chosen for the 420 bulb turbine units, each of 16.2 MW capacity and 10 m in diameter. The powerhouse, 10.5 km long, would be formed by

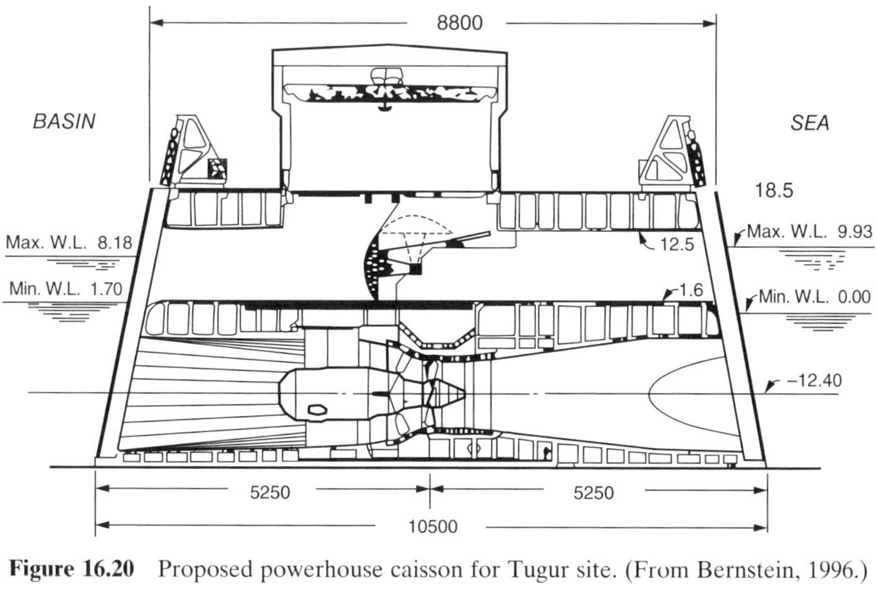

Figure 16.20 Proposed powerhouse caisson for Tugur site. (From Bernstein, 1996.)

105, four-unit caissons of the typical cross section of Fig. 16.20. The powerhouse would also include three fishway caissons designed for the purpose. The remainder of the barrage connecting the powerhouse to the shores would be rock-fill embankments with 4:3 slopes secured underwater by tetrahedrons and above water by slabs. At the end of the eighth year of construction, when the dam had been erected and the caissons installed, sufficient head would be available for the operation of those turbines that had been installed and, thereafter, as generator units were installed, they could be placed in operation so that before the plant is completed a return on the investment would be initiated. It is considered that, early in the next century, a Tugur tidal-electric development would form an integral and essential component of the power system required in the Far East and Yakut region, which will also contain a large component of hydroelectric power and yield a substantially less energy cost than the thermal alternative.

16.11.4 Kolskaya Tidal Power Scheme

It is realized that such large developments as the Mezen and Tugur plants will require further extensive investigations, particularly with regard to the ecological aspects, as well as substantial international cooperation; complex problems stemming from the geographical locations and hostile climatic environments can also be expected. Although the experimental Kislaya Guba project met its basic objective of testing construction methods and materials for the development of the large potential sites, it would be classified only as a mini-tidal-electric station at 0.4 MW as opposed to a large Tugur development of 6800 MW. Bernstein (1992) suggested that an intermediate-sized plant be constructed to test a 10 m diameter runner with different modifications, such as connections to stepup gears, variable frequency of rotation, as well as double-effect operation and other construction, technological, and ecological solutions. For this purpose, the construction of a pilot commercial plant at Kolskaya on the Murmansk coast of the Barents Sea consisting of one, 16.2-MW unit has been proposed. The area of the bay at 5 km^2 and the average tidal range of 2.36 m have been considered sufficient for the desired trials. Although the cost of constructing the Kolskaya barrage may be high with respect to the energy produced as a result of the low tidal range available, Bernstein considered that its cost would be justified by the significant reductions that would be made possible in the cost of the proposed Tugur and Mezen tidal-electric plants (refer to Figs. 16.18 and 16.19).

16.12 UNITED KINGDOM

In his book, A. C. Baker (1991) has brought together the studies carried out in the United Kingdom during the 1970s and up to 1989.

16.12.1 Severn Estuary

Interest in harnessing the high tides of the Severn Estuary, which have a mean range of 8.5 m, began early in the twentieth century. Government interest in the possibility of cxploiting the tidal energy in the Severn Estuary commenced with the setting up of an official working group in 1925. Its 1933 report proposed a development of 800-MW capacity but could not make a clear, economic case for development and the proposal was not pursued. By the mid-1970s, the problem of ensuring a sufficient and reliable energy supply for the long-term future stimulated further interest. Based on a broad initial assessment of the technical problems, the Department of Energy established the Severn Barrage Committee in 1978 to advise the government on the desirability of a tidal scheme in the Severn Estuary.

The committee's report of 1981 recommended the single-basin, ebb generation scheme on the alignment between Lavernock Point on the Welsh shore and Brean Down on the English shore, known as the Cardiff–Weston line (refer to Fig. 16.21), be selected and that a further phase of work be undertaken on this scheme, known as the Inner Barrage. More intensive study was also recommended to establish the acceptability of this scheme with regard to

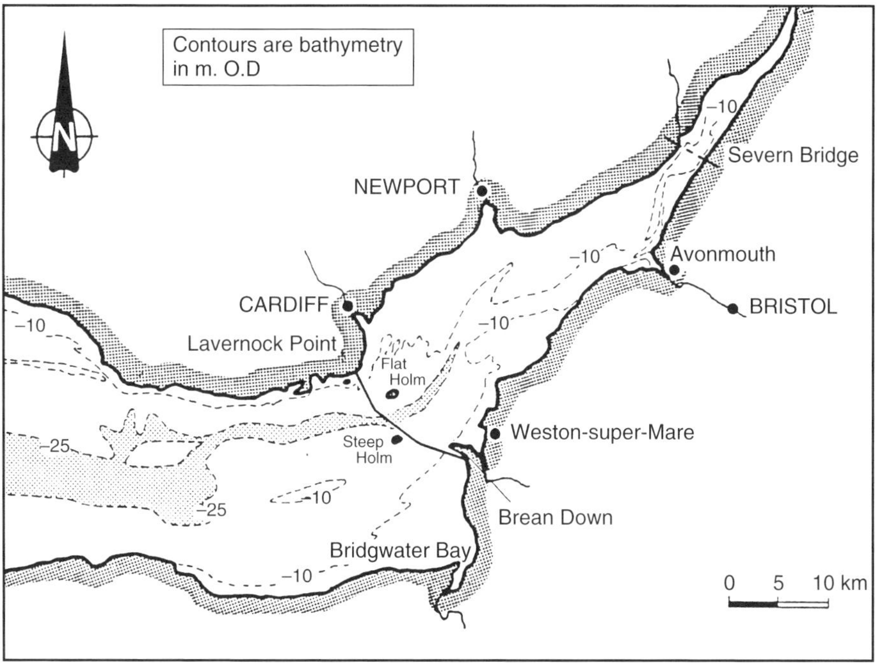

Figure 16.21 Inner barrage alignment—Severn River Estuary. (From Severn Tidal Power Group, 1989.)

environmental, social, and industrial factors coupled with preliminary design and further economic evaluation.

In 1984, a consortium of six construction companies in the United Kingdom formed the Severn Tidal Power Group (STPG) with the aim of furthering the project. The 1986 report of that group confirmed the choice of the Inner Barrage with minor modifications and set out the results of its assessment on the commercial viability of the project, built and operated in the private sector. From 1987 to 1989, the group undertook a "development" study of the Inner Barrage with the work being funded equally by the group, the Central Electricity Generating Board, and the Deparment of Energy. The study examined further the design and construction of the barrage so as to reduce the uncertainties relating to its cost and performance in order to determine, in greater depth, its regional and environmental effects and to reassess its economic viability as well as to define further work needed before a decision could be taken to proceed to construction. The main report on this study was published by STPG in 1989 and detailed reports on various aspects of the study, in 1990. The study confirmed that the Inner Barrage (Fig. 16.22) is a feasible project, well within the scope of existing technology, and could have a very long life (120 years) with reasonable maintenance. The main design characteristics for the Inner Barrage, as determined by this study, are set out in Table 16.7 and the layout of the project is shown in Fig. 16.22.

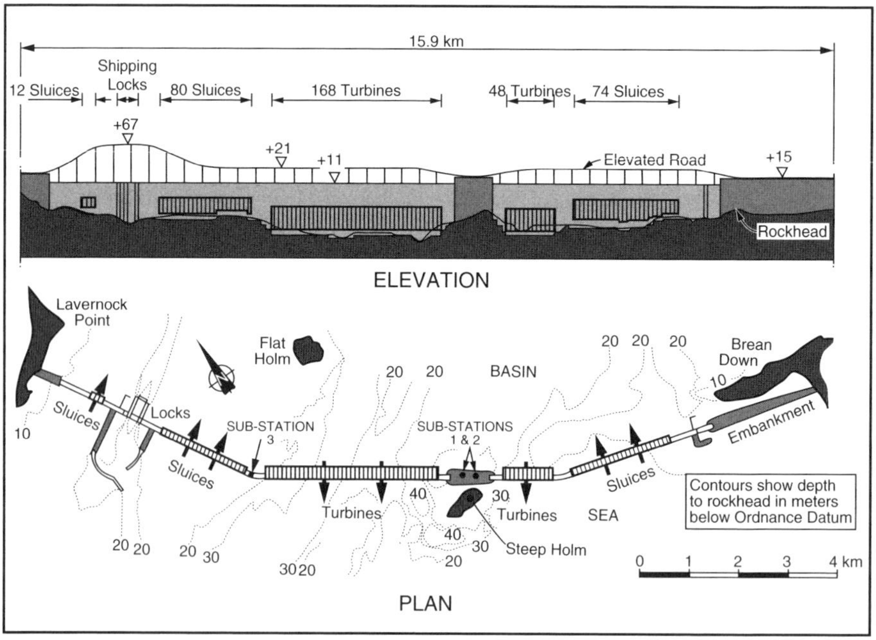

Figure 16.22 Proposed layout of the Severn Barrage project. (From Severn Tidal Power group, 1989.)

TABLE 16.7 Main Design Characteristics for the Inner Barrage, Severn Estuary

Number of turbine generators	216
Diameter of turbines	9.0 m
Operating speed of turbines	50 r.p.m.
Turbine generator rating	40 MW
Installed capacity	8640 MW
Number of sluices (various sizes)	166
Total clear area of sluice passages	35,000 m^2
Average annual energy output	17 TWh
Operational mode: ebb generation with flood pumping	
Total length of barrage	15.9 km
Including: powerhouse caissons	4.3 km
sluice caissons	4.1 km
other caissons	3.9 km
embankments	3.6 km
Area of enclosed basin at mean sea level	480 km^2

The study concluded that the annual output of about 17 TWh would amount to 7 percent of the then current electricity consumption in England and Wales and displace the burning of 8 million tonnes of coal equivalent of fossil fuel annually, which would help the strategy to minimize the threat of the greenhouse effect by reducing carbon dioxide emissions by up to 17.6 million tonnes per year and would not cause acid rain. The report pointed out that if renewable energy resources are to be utilized to increase diversification of electricity generation and to reduce pollution, the Severn tidal-electric project remains the largest single project that could make a significant contribution on a reasonable time scale and, when completed, would represent an insurance against escalation of fuel prices. The report also noted that there was still considerable work to be carried out in order to complete a full environmental assessment.

16.12.2 Other Potential Sites

In its report, the Ad Hoc Working Party on Tidal Power (1978) listed the major tidal power sites in the United Kingdom, in addition to the Severn Estuary as the Wash, Solway Firth, and Morecambe Bay (refer to Fig. 16.23). However, during the past two decades none of the latter sites has received any special mention in the literature. On the other hand, studies were initiated to assess the feasibility of exploiting the tidal energy potential in other estuaries, such as Mersey (700 MW), Conwy (33 MW), Wyre (60 MW), Loughor (5 MW), and Duddon (100 MW). Shaw (1997) outlines the work that has been carried out on these estuaries to date. Proposed developments in the Conwy and Wyre estuaries could provide an annual output of 60 and 130 GWh, respectively (Matthews and Young, 1992a,b). Feasibility studies would be required to confirm the barrage designs and to develop the necessary background on

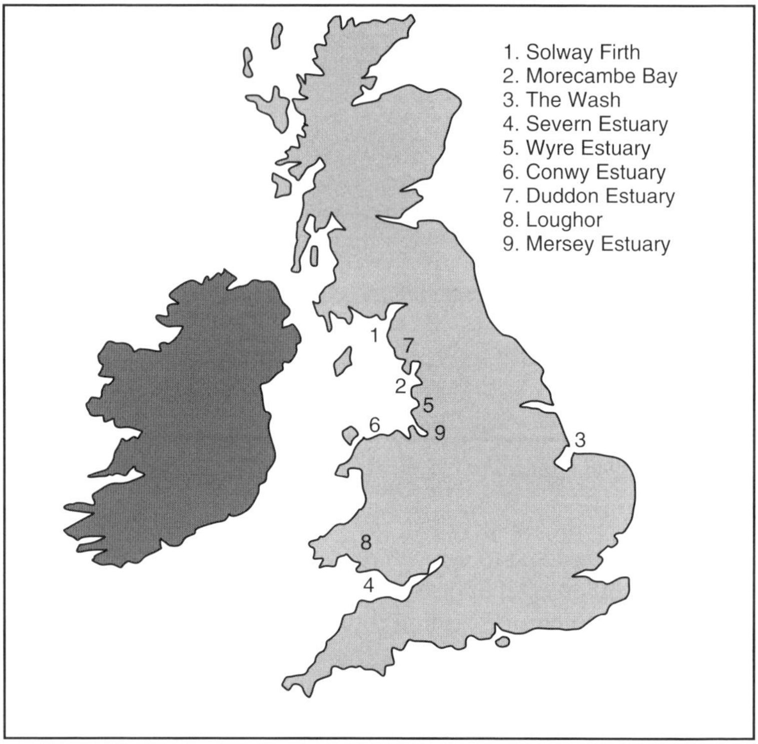

Figure 16.23 Major tidal power sites in the United Kingdom.

the ecological and other socioeconomic aspects related to such projects (Shaw, 1997a,b).

A feasibility study (stage III) of a Mersey Barrage was initiated in 1986 and completed in 1991 (Conroy et al., 1992). The main design characteristics of the scheme are given in Table 16.8. Studies are still being carried out to address, in more detail, the concerns that have been expressed by shipping and environmental groups regarding this project and to develop firm estimates of the construction cost and energy output as well as an appropriate financing structure. The studies to date have shown the technical feasibility of the project, although its economic feasibility has not yet been indicated (Wilson, 1992).

16.13 UNITED STATES

16.13.1 Cobscook Bay, Maine

Studies have been carried out by the U.S. Corps of Engineers to determine the feasibility of exploiting the energy of the tides in Cobscook Bay, which lies adjacent to Passamaquoddy Bay but is wholly within the United States. Since

TABLE 16.8 Main Characteristics of the Mersey Barrage

Length between river banks	1900 m
Mode of operation	Ebb generation with pumping
Number of turbines	28, each 8 m in diameter
Number of sluices (rising sector gates)	46, each 17 m wide
Turbine generator rating	25 MW
Installed capacity	700 MW
Average annual energy output	1.4 TWh

Source: Adapted from Wilson (1992).

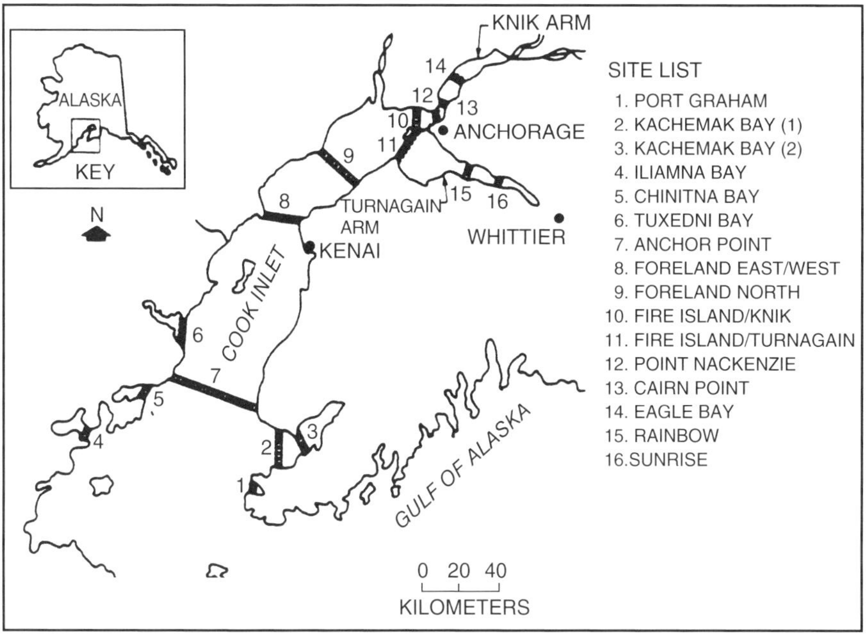

Figure 16.24 Potential tidal power sites—Cook Inlet.

the bay contains many islands (refer to Fig. 16.12), several projects have been proposed for the damming off of various areas within the bay for the installation of an ebb generation, single-effect tidal-electric plant. In view of the number of islands, it would also be possible to create a double-basin scheme. Proposed installed capacities have ranged from 5 to 180 MW. Cobscook Bay has a mean tidal range of 5.4 m and an area of about 98 km^2.

16.13.2 Cook Inlet, Alaska

The mean tidal range in the Knik Arm and Turnagain Arm of Cook Inlet, Alaska (Fig. 16.24), reaches about 8 m. A preliminary assessment of the tidal power potentials and characteristics, undertaken for the governor of the State of Alaska, was completed in 1981. It was concluded (Acres American Inc., 1981) that studies should proceed to evaluate those potential industries that would have a comparative advantage in association with a Cook Inlet tidal-energy source and for the preparation of a conceptual development plan for a site above Eagle Bay in Knik Arm.

It was considered that a tidal electric plant at this location could have an estimated capaciy of 1440 MW from 60 turbogenerator sets and 36 sluices with an annual energy output of about 4037 GWh (Warnock, 1987). The barrage would be formed by prefabricated powerhouse and sluiceway caissons. The preliminary assessment also concluded that tidal power development at Eagle Bay need not be dependent on the encouragement of industrial growth in the region since this site could be developed in a manner consistent with the most likely energy demand growth in the region, provided the energy could be retimed economically. A major field program would be required to remove or reduce the uncertainties arising from the limited database and environmental concerns.

APPENDIX A

TIDAL GENERATION OPTIMIZATION MODELS*

SINGLE-TIDE OPTIMIZATION MODEL (STOM)

The derivation of the model is based on the simulation of the levels of the impounding reservoir or tidal basin over time, $y(t)$, using the following equations:

$$H = y(t) - h(t) \tag{A.1}$$

$$Q = -A(y)\frac{dy}{dt} \tag{A.2}$$

$$Q = Q(H) \tag{A.3}$$

where h_t is the sea level, $Q(H)$ the turbine or sluice discharge characteristic, and $A(y)$ the tidal basin area as a function of level. Since the tidal basin surface area is known at all water levels, Eq. (A.3) is a continuity equation, equating $Q\,dt$, the aggregate volume of water discharged through turbines and sluices, to $A(y)\,dy$, the reduction of storage, that is, changes in tidal basin water level. As implied in Chapter 3, the characteristics of the basin can have considerable impact on energy production so that the relationship between

* The model derivations have been abstracted from the reports of the Bay of Fundy Tidal Power Review Board and Management Committee, 1977.

basin water level and volume, which can be determined at small cost, should be as accurate as possible.

If $Q(H)$, the discharge characteristic, is *deterministic*, then the foregoing equations may be combined into a single differential equation:

$$\frac{dy}{dt} = \frac{-Q[y(t) - h(t)]}{A(y)} \tag{A.4}$$

The numerical value of the right side can be computed for any values of t and y. The function $y(t)$ is, therefore, a solution of a differential equation of the form

$$\frac{dy}{dt} = f\,bg$$

This equation can be solved using the Runge–Kutta method, that is, a step-by-step numerical integration. Computations must start at a given point when the level y_0 and the time t_0 are known. Level y_1 at time $t_1 = t_0 + \Delta t$ is computed from the following equations:

$$k_1 = \Delta t f\,(t_0, y_0) \tag{A.5}$$

$$k_2 = \Delta t f\left(t_0 + \frac{\Delta t}{3}, y_0 + \frac{k_1}{3}\right) \tag{A.6}$$

$$k_3 = \Delta t f\left(t_0 + \frac{2}{3}\Delta t, y_0 - \frac{k_1}{3} + k_2\right) \tag{A.7}$$

$$k_4 = \Delta t f\,(t_0 + \Delta t, y_0 + k_1 - k_2 + k_3) \tag{A.8}$$

$$y_1 = y_0 + \frac{1}{8}(k_1 + 3k_2 + 3k_3 + k_4) \tag{A.9}$$

The problem execution can be carried out either forward or backward and the step-size or integration interval is automatically adjusted to specified error bounds. Therefore, a starting level can be computed knowing the finishing conditions, or vice versa. Generally, a time step of 5 or 10 min will be found to be a satisfactory compromise between accuracy and execution time. Under certain conditions, such as at the end of the turbining period, a time step of 1 min may be found to be necessary. To determine the relationship between head and discharge for each time interval, appropriate turbogenerator performance curves must be used.

When $Q(H)$ is not deterministic only with respect to head, such as when a variable-pitch blade turbine is involved, a more complex formulation is required. In order to optimize the energy production during a period of generation (t_1, t_2), the discharge during that period must be selected so that

$$E = \sum_{t_1}^{t_2} p(t)N(Q, H)dt \tag{A.10}$$

is a maximum where E is the revenue for the period, N is the power output varying with Q *and* H, and $p(t)$ is the monetary value per kilowatt hour of energy varying with time.

Using Eqs. (A.1) and (A.2),

$$E = \sum_{t_1}^{t_2} p(t)N[-A(y)y', y - h(t)]dt$$

or

$$E = \sum_{t_1}^{t_2} f(t, y, y')dt$$

where y' corresponds to the first derivative dy/dt.

This equation represents a classical problem of finding a function $y(t)$ in order to maximize E. Using the calculus of variations, it may be shown that $y(t)$ must be the solution of Euler's partial differential equation:

$$\frac{\partial f}{\partial y} - \frac{d}{dt}\left(\frac{\partial f}{\partial y'}\right) = 0 \tag{A.11}$$

with transversality conditions

$$\left(f - y'\frac{\partial f}{\partial y'}\right)\partial t - \left(\frac{\partial f}{\partial y'}\right)y = 0 \tag{A.12}$$

In the present case,

$$\frac{\partial f}{\partial y'} = p(t)\left[-y'\frac{dA(y)}{dy}\frac{\partial N}{\partial Q} + \frac{\partial N}{\partial H}\right]$$

and

$$\frac{\partial f}{\partial y'} = -p(t)A(y)\frac{\partial N}{\partial Q}$$

Hence,

$$p(t)\frac{\partial N}{\partial H} + p(t)A(y)\frac{d}{dt}\left(\frac{\partial N}{\partial Q}\right) + A(y)\frac{\partial N}{\partial Q}\frac{dp}{dt} = 0 \tag{A.13}$$

Equation (A.13) is the fundamental equation that defines the local conditions for an optimal relationship between the discharge, basin level, and time during the generation period. It can be evaluated in finite difference form and solved numerically in the model.

In addition to this equation, the transversality conditions of Eq. (A.12) provide some guidance on the starting and finishing conditions of a period of generation. In this case,

$$\left(N - Q\frac{\partial N}{\partial Q}\right)\partial t - A\left(\frac{\partial N}{\partial Q}\right)\partial y = 0$$

If the starting level of the basin is given, then $\partial y = 0$ and the turbines must commence operation under the condition

$$N - Q\left(\frac{\partial N}{\partial Q}\right) = 0$$

or, by introducing the efficiency,

$$\eta(Q, H) = \frac{N(Q, H)}{\rho g Q H}$$

$$\frac{\partial \eta}{\partial Q} = 0$$

Hence, turbining must start (or end) with a discharge giving maximum efficiency. Although this condition does not fully define the starting conditions, the number of possible routes through the turbine hill chart is reduced.

Equation (A.13) is solved numerically during the generating period. Conditions are defined at the start of the computation by the optimization routine as t_0, y_0, Q_0, and H_0. New values are computed at $t_1 = t_0 = \Delta t$ using the finite difference form of Eq. (A.13):

$$p(t_0)\frac{\partial N}{\partial H}\left[\frac{Q_0 + Q_1}{2}, \frac{H_0 + H_1}{2}\right] +$$
$$\frac{p(t_0)}{\Delta t} \times A\left[\frac{y_0 + y_1}{2}\right] \times \left[\frac{\partial N}{\partial Q}(Q_1, H_1)\frac{\partial N}{\partial Q}(Q_0, H_0)\right] +$$
$$A\frac{(y_0 + y_1)}{2} \times p\frac{(t_0 + t_1)}{\Delta t} \times \left[\frac{\partial N}{\partial Q}\left(\frac{Q_0 + Q_1}{2}\right), \left(\frac{H_0 + H_1}{2}\right)\right] = 0 \qquad \text{(A.14)}$$

Equations (A.1) and (A.2) provide the first estimate of y_1 assuming $Q_1 = Q_0$, so that

$$y_1 = y_0 - \Delta t\left[\frac{Q_0}{A(y_0)}\right] \qquad \text{(A.15)}$$

$$H_1 = y_1 - h(t_1) \tag{A.16}$$

Having obtained a value for y_1 and H_1, Eq. (A.14) becomes an equation in Q_1. A solution is found by searching for Q_1 from Q_0. Improved values of y_1 and H_1 are then obtained by a second use of Eqs. (A.15) and (A.16), where Q_0 is replaced by $(Q_0 + Q_1)/2$.

The computation should properly be started using the new estimates of y_1 and H_1, thereby initiating an iterative process. However, such an approach is very time consuming and an alternative may be satisfactory. A time step of 2 min could be used and, by computing an initial y_1 by Eqs. (A.15) and (A.16) and computing Q_1 from Eq. (A.14), the new value of Q_1 is used to recompute y_1 from Eqs. (A.15) and (A.16). At each 10-min interval, closing errors are balanced and the solution smoothed by recomputing the reservoir level using the average discharge over the last five steps.

Values of N, $\partial N/\partial Q$, and $\partial N/\partial H$ were computed from an equation fitted to the hill chart. For the Fundy studies, a step length of 2.5 m^3/s was used in the search for Q_1. The equations may be used with either a positive or negative time step to compute forward or backward in time.

Equation (A.14) is no longer valid when the solution implies operation above the generator limiting capacity. In order to limit the turbine output as the head increases, the turbine must be throttled to follow the line of limiting generator output (refer to Fig. 9.8*b*). Figure A.1 illustrates a typical path, *VWXYZ*, followed through the hill chart.

At the intersection of the path *VW* with the limiting power curve, that is, at point *W*, the discharge becomes deterministic with head and the Runge–Kutta method is used as long as operation continues at generator limiting output.

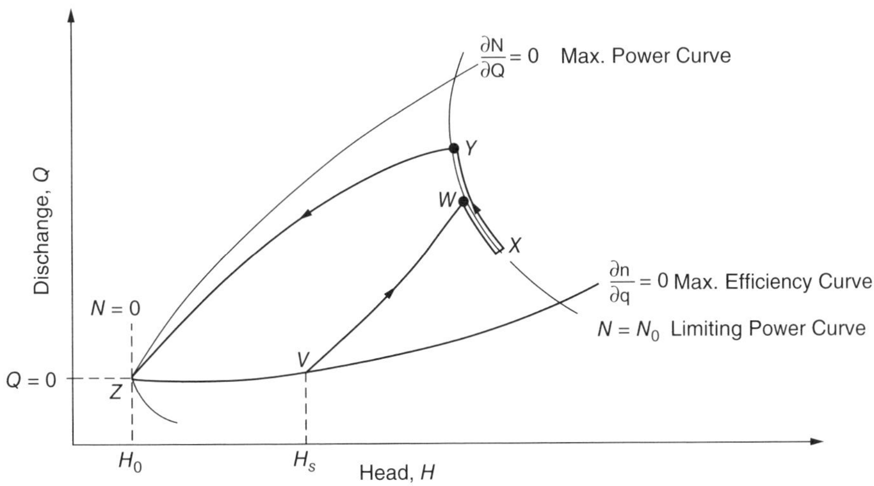

Figure A.1. Typical operation for variable-pitch blade turbine.

According to Gibrat (1966), the turbines must operate along the limiting power curve until the following equation is satisfied:

$$\sum_{W}^{Y} \frac{p(t)\left(\dfrac{\partial N}{\partial H}\right) + A(y)\dfrac{d}{dt}\left(p(t)\dfrac{\partial N}{\partial Q}\right)dt}{p(t)A(y)\dfrac{\partial N}{\partial Q}} = 0 \qquad \text{(A.17)}$$

This integral is computed from point W until it is satisfied, point Y. A finite difference form of Eq. (A.17), which is similar to Eq. (A.14), is evaluated numerically in the simulation models. Computation continues using Eqs. (A.15), (A.16), and (A.14) along the path YZ.

ONE-CYCLE DYNAMIC PROGRAMMING MODEL

The *one-cycle dynamic programming model* (OCDPM) is also based on Eqs. (A.1), (A.2), and (A.3).

The program logic introduces a function G that is defined as the maximum benefit that can be obtained by operation of the tidal power plant during a single tidal cycle when the starting and finishing levels are known. Over a finite number of cycles, there can be only one choice of basin starting level for each cycle that produces an overall maximum, provided that the operation is always optimized during each individual cycle.

The *dynamic programming* technique is used to select the reservoir level at the beginning of each generating period. If the period being considered is divided in steps $S_1, \ldots, S_i, \ldots, S_n$ and the reservoir levels at the beginning of each step are designated $y_1, \ldots, y_i, \ldots, y_n$, then there is only one optimal route producing maximum energy provided, of course, that generation is optimum during each step, that is, between levels y_i and y_{i+1}. The maximum benefit during step S_i can be represented by the function $G_i(y_i, y_{i+1})$, with the given starting and finishing conditions of y_1 and y_{i+1}, respectively. For the case where the value of energy is constant, the maximum benefit is identical to maximum energy production. The problem reduces to the determination of the optimal route of levels, y_i, in order to maximize the sum of n partial benefits during n steps. The best operation during one cycle and between two predetermined levels, y_i and y_2 (refer Fig. A.2), is established as follows for single-effect, fixed-blade runners with no pumping:

1. The lowest possible level (at the end of turbining), y_0, from which the basin can be refilled to level y_2 is computed. Since y_0 is a function of y_2 only, it may be computed by solving backward the refilling equations at maximum capacity. Because of the minimum operating head of turbines, the reservoir can only refill to y_2, provided that turbining ends on the curve ABC of Fig. A.2*b*.

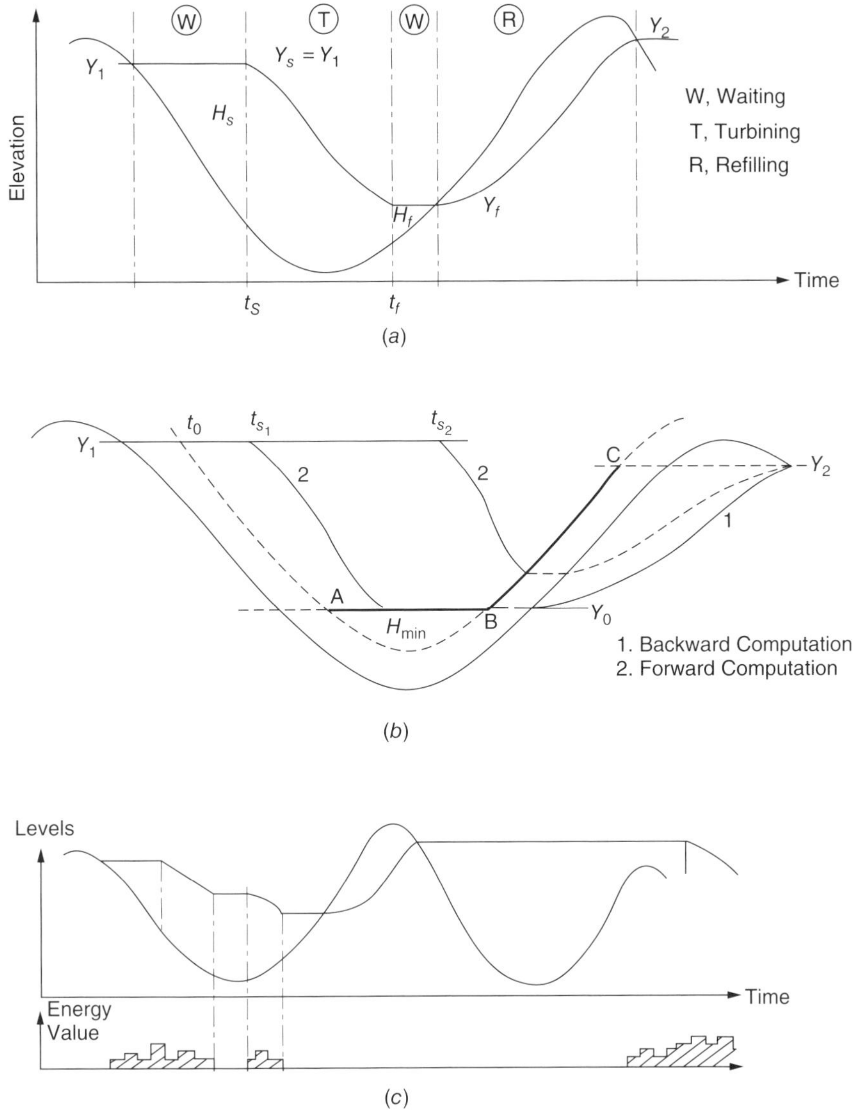

Figure A.2 Single-effect operation with one-cycle dynamic programming model. (*a*) single effect without pumping, (*b*) single effect suboptimization, and (*c*) single effect with zero energy value.

2. For a given starting time, t_s, the reservoir level during turbining is computed by solving forward the corresponding differential equation. Turbining would end either when y_0 is reached (t_{s1}) or when the minimum operating head is reached (t_{s2}). The energy production and benefit are simultaneously computed. Note that the benefit is a function of only one variable, the starting time, t_s.

3. The maximum benefit, which is stored as $G_i(y_1, y_2)$, is found directly by varying t_s according to a given time step, t, starting from a minimum time, t_0; the time step, t, should be reduced in the region of maximum benefit.
4. If the energy value is zero for a period of time, the turbines are shut down until it again has a value. The model program must recognize this situation otherwise energy production could, on occasion, occur at a time of zero energy value. If the energy value were zero for a complete cycle, the reservoir level would be maintained to enhance the operation during the subsequent cycle (Fig. A.2*c*).
5. The starting levels are optimized by solving the following equation:

$$R_{i-1}(y_1) = \max\ y_2\,[G_{i-1}(y_1, y_2) + R_i(y_2)] \tag{A.18}$$

for y_1 with steps of Δy varying between two extreme values. R_i is the total benefit (energy production) between the present time, i, when the reservoir is at level y_2, and the end of the time period. G_{i-1} is the benefit obtained by operating the reservoir between levels y_1 and y_2 during time interval $i - 1$, and the sum of G_{i-1} and R_i is maximized to select the optimum route between a given level y_1 to the appropriate level y_2 at he beginning of time stage i.
6. The lowest level of y_1 is that level to which the reservoir would refill from a level equal to the low tide level of the previous cycle plus minimum operating head, that is, the lowest possible level at the end of the previous turbining period.
7. The highest value of y_1 is set equal to the lesser of either the greater of the current and preceding high tide level or a given upper limiting level. This upper limit assumes that, although it may be desirable not to generate during one cycle, it would never be optimal to skip two successive cycles.
8. Level y_1 is varied in descending order, and for each such level y_2 is varied in ascending order. Limits of variation of y_2 are the limits of y_1 for the next cycle.
9. The computation would not necessarily be repeated for each new value of y_2 because the optimal operation remains the same so long as the limiting level y_0, associated with a new y_2, is less than the finishing level at the end of turbining.
10. When the full series of tides has been explored, the results are reviewed to establish the optimum route through the series from a given initial level in the first cycle. A single computation of the draw-down or turbining period for each cycle is repeated to provide the final results, that is, the energy production during each hour of the series of tides.

Maximum benefit $G_{i-1}(y_1, y_2)$ can be readily estimated when the operation (turbining, pumping, sluicing) between levels y_1 and y_2 remains constant during the time interval S_i. This is true when the step is short. The new version of the La Rance tidal power plant operation model (Merlin et al., 1980) still uses a 10-min time step, an interval made necessary by the speed of variation of the sea level, which can reach 3 m/h. In the 1970s, the computer execution time required when using such a step was very expensive for the type of study being undertaken in the Fundy reassessment. Therefore, for those studies, the time step used coincided with one tidal cycle on single-effect generation models and with one-half cycle on double-effect models. For single-effect models, in particular, the OCDPM achieved an important reduction in computation costs with a negligible loss of accuracy.

GLOSSARY OF TERMS

Armoring refers to the protection of surface layers of an embankment by rock or specifically designed concrete units such as dolosse.

Barrier or **Barrage** is a permanent structure created within a coastal waterway to control both water level and discharge in an estuary by means of gates; barrage is a relatively low structure that controls water in the estuary generally within its natural range and often does not need to be as impervious as a river dam.

Barring is a term used in the thermal generation industry to indicate a turbine that is in a state of readiness for load acceptance. When "on barring," a turbine rotor is turning slowly under the supplementary power of a barring motor or barring gear. The purpose of barring is twofold: When not is use, the slow rotation of the rotor prevents the oil film protecting the bearings from being squeezed out by the static weight of the unit; after full-load rejection, barring is used to minimize heat shock and ensure uniform cooling. The term originates from the original method of using crowbars inserted into holes in wheels that were used for turning.

Base Load is the minimum load imposed on an electrical system over a specified period of time. A base-loaded generation unit is dedicated to meeting a more-or-less continuous electrical demand.

Bathymetry is the depth configuration of the ocean floor.

Elements of Tidal-Electric Engineering. By Robert H. Clark

Benefit/Cost Ratio as used in this text is the ratio of the present worth, net benefits of savings (viz. net of operating costs) to the electrical utility system, to the present worth of cost and maintenance of the tidal-electric development.

Blank (or **Blind) Caisson** is one with no openings for water flow and may be used instead of embankment.

Caisson is a watertight vessel, buoyant under its own weight, that can be floated into a predetermined location and then sunk in place as an element by flooding; for the proposed tidal power technology skeletal powerhouse and sluiceway modules could be constructed offsite and, after accurate placement on the barrage alignment completed thereafter in situ or the turbogenerator and gates could be fitted in the respective caissons before floating them to the site.

Capacity Contribution or **Firm Capacity** of a tidal power scheme is measured by the total installed capacity of other conventional generating plants that would otherwise have to be constructed to meet the same electrical power demand with the same loss of load probability.

Cofferdam is a temporary structure, often of steel-sheet pilings, erected to exclude water from an area that would normally be submerged to facilitate the construction of foundations under dry conditions.

Constant Dollar values are expressed in unit prices considered to prevail at a particular point in time, usually at a specific time within the feasibility study.

Conventional Generating Stations in the context of this text employ established technologies and are river hydroelectric, coal, oil, gas, and nuclear power stations.

Coriolis Effect causes a body moving over Earth to be deflected from a straight path relative to the Earth's surface because Earth is rotating; this is termed the Coriolis effect (after the French scientist who first described it) and in the northern hemisphere is always to the right; thus, for example, water in the Bay of Fundy tends to pile up along the New Brunswick, Canada, shoreline.

Current Dollar values are expressed in unit prices considered to prevail at the actual time that expenditures are incurred or revenues received; future current dollar values can be estimated by increasing the constant dollar prices by price inflation factors pertinent to the period and subject under consideration.

Cycling of an electrical generating plant describes the periodic running up and down of its output; for some types of plant this can result in penalties to thermal efficiency, maintenance requirements, and plant lifetime.

D_{50} (in millimeters or meters) is the median size of particle acceptable for quarry-run rockfill material for the construction of dikes.

Diurnal Constituent of a tidal wave form is one that has a periodicity that is close to the duration of a solar day.

Diversity of Supply in an electrical generating system relates to the number of types of generating plant and fuels used; with increased diversity, the risk of failure to meet electricity demand, such as through plant-type faults or fuel shortage, is reduced.

Double-Effect operation refers to the generation of tidal power on both the ebb and flood tides.

Double-Regulated turbine is one with two separate methods of regulating the water flow and hence power output, viz. a turbine with adjustable guide or distributor vanes (also known as wicket gates) and adjustable runner blades.

Drydock or **Graving Dock** is a docking facility accessible to the ocean in which the powerhouse and sluiceway module caissons would be constructed so that, after fabrication, they could be floated in place by flooding the dock and then towed out for placement in the barrage.

Ebb Generation is the mode of tidal power generation in which water passes through the turbines in the same direction as the ebb tide, i.e., from the basin to the sea.

Ecosystem is a community of organisms and their environment.

Energy Potential is used to describe the amount of energy that could theoretically be extracted from the tides at a given barrage site by using very large numbers of turbines.

Estuary is that segment of a river influenced by the tide of the water body into which it flows.

Fetch is a measure of the distance over water in the direction of a wind current over which wind-induced waves are assumed to build in amplitude.

Firm Capacity is the capacity of a given generation unit or system of units that is expected to be available, when required, to meet a load demand.

Flood Generation is a mode of tidal power operation in which water passes through the turbines in the same direction as the flood tide, i.e., from the sea to the basin.

Foreshore is the zone between low water and high water.

Generator Rating, or rated electrical output, is the normal maximum output.

Geostrophic Coefficient, γ, is used in the tidal regime modeling analyses and is the effective component of the Coriolis force in the horizontal plane; algebraically, it is defined by the equation $\gamma = 2\Omega \sin\theta$ wherein Ω is the angular rotational velocity of Earth (7.27×10^{-5} rad/s) and θ is the latitude of the location of interest.

Harmonic Analysis is a procedure used to compute, from observational data, the individual amplitudes and phases of significant tidal constituents that make up the complex form of a tidal wave.

HHWLT, higher high water large tides; this level is the highest that can be predicted from the available tidal constitutent information.

Ice (Cake Ice) is composed of individual masses of ice that may combine to form part of an extensive pack; cake ice aggregates in bays and estuaries and increases in thickness due to the periodic accretion of layers of frozen muddy water and sediment as it grounds at low tide.

Ice (Sheet Ice) is a regular ice cover that is free of deformities (ridging, rafting, and overlapping) caused by pressure.

Ice (Shorefast Ice) remains attached to shores and banks in estuaries and embayments and can build up to great thicknesses where steep gradients and rocky terrain prevail in the intertidal zone.

Intertidal Area is the zone of the seabed between low water and high water.

LLWLT, lower low water large tides; this level is the lowest that can be predicted from the available tidal constituent information.

Levelized Cost (of energy) is a weighted average cost of energy production over a specified period of time determined by utilizing price discounting procedures (see Present Worth): In this text, it is the present value of the production cost of the energy that is displaced by tidal-generated energy divided by the amount of energy displaced discounted at the same rate. In algebraic terms it may be defined by

$$\text{Levelized cost} = \sum_{i=1}^{i=n} \frac{C_i}{b+dg} \Big/ \sum_{i=1}^{i=n} \frac{E_i}{b+dg}$$

where E_i = energy displaced in year, i
C_i = production cost of the energy displaced in year, i
d = discount rate.

(Note that $C_i = E_i \times m_i$, where m_i is the unit production cost of the displaced energy.)

Load Duration Curve is a (graphical) representation of the relationship between time and the electrical demand of the system, i.e., it is a plot of load demand for each time interval under consideration.

Load Factor is the ratio of the average demand to the maximum or peak demand over a specified time period.

Lunar (Tidal) Day is the duration of a single rotation of Earth on its axis relative to the moon. Its average is about 24 h and 50 min of mean solar time.

Mathematical Model is used synonomously with numerical model.

Merit Order is a list of generating plant in ascending order of avoidable generating cost.

Multiplier Effect is a term used to describe the phenomenon whereby the wealth injected into a community as a result of a large project indirectly leads to new jobs in the locality other than those resulting directly from the project.

Numerical Model is a mathematical simulation of a real situation or a real process or effect typically involving complex variations of state in time or space, or both.

Overburden is the overlying stratum of soil relative to an elevation of interest, frequently used to refer to all soil materials overlying the bedrock and often used to include all unconsolidated and weak materials lying above a proposed foundation level.

Peak Demand is the maximum capacity demand on an electrical supply system over a specified period of time.

Piping is a form of a failure in a permeable granular structure that is caused by either scour or subsurface erosion. It occurs, e.g., near the downstream toe of a water-retaining embankment, when the pressure gradient resulting from seepage is sufficient to induce motion in the soil particles within which the seepage is taking place.

Present Worth or Value is the measure of value ascribed to investments and revenues in this text; it represents the value, in a base year, of a dollar in any other year or a series of dollar values over a number of years, after adjustment for the effects of price inflation; alternative descriptions for present worth commonly used are "present value" and "discounted value".

Resonance is the enhanced amplification of a tide, the frequency of which is close to the natural frequency of oscillation of water in an estuary.

Semidiurnal Constituents of the tidal wave form are those that exhibit a periodicity that is close to one-half of the duration of a solar day; in the Bay of Fundy, Canada, e.g., the semidiurnal constituents are much greater than the diurnal constituents.

Significant Wave Height is computed as the mean of the largest one-third of the waves for each storm condition evaluated.

Single-Effect operation of a tidal power development in this text (and generally) refers to generation only on the ebb tide.

Spectral Analysis is a method of analyzing a complex wave form expressed as a time series to determine which frequencies display the maximum energy.

Synchronized. The output of a turbogenerator is said to be synchronized with the system grid when the waveform of its alternating electrical output is locked in phase with the grid.

System Cost is the total cost of meeting electricity demand and is the sum of various elements, such as capital, fuel, maintenance, overhead, etc.

System Security refers to the degree of assurance with which the generating system can meet the annual peak in electrical demand; such assurance is obtained by installing a generating plant with a total rated capacity in excess of this peak to allow for plant availability in winter, load estimating errors, and weather of unusual severity.

Thermal Cycling refers to a method of utilizing a thermal generation plant in the midpart of the load–duration curve whereby the plant output is continually varied to meet changes in the pattern of demand.

Tide is the effect on the waters of Earth by the gravitational attraction of celestial bodies, principally the moon and to a lesser extent the sun, whereby they tend to become elevated at a point beneath the moon (and at the opposite side of Earth's circumference to this location) and hence depressed at points midway between these locations; in the Bay of Fundy, Canada, where the semidiurnal constituents of gravitational attraction are dominant, the result is a twice daily (lunar day) inflow and outflow of a tidal wave relative to the coastline.

Tide (Ebb) is the efflux of tidal waters away from a coastline.

Tide (Flood) is the influx or rising stage of tidal movement toward a coastline.

Tide (Neap) is a lower than average high tide that occurs when the moon is in the first or third quarter and its graviational attraction is opposed to that of the sun.

Tide (Spring) is a higher than average tide that occurs when the moon is new or full and its gravitational attraction is complementary to that of the sun.

Tidal Amplitude is one-half of the tidal range.

Tidal Constituents are the individual, sinusoidal waveforms that collectively produce a tide. The principal constituents identified in the text are:

M2—the principal lunar semidiurnal constituent (and by far the largest tidal component) with a speed of 28.984 degrees/mean solar hour (°/msh).

N2—the larger lunar elliptic semidiurnal component with a speed of 28.440°/msh.

L2—the smaller lunar elliptic semidiurnal component with a speed of 29.528°/msh.

K2—a lunar declinational semidiurnal component with a speed of 30.082°/msh.

S2—the principal solar semidiurnal component with a speed of 30.000°/msh.

K1—a lunar declinational diurnal constituent with a speed of 15.041°/msh.

O_1—lunar declinational diurnal constituent with a speed of 13.943°/msh.

μ_2—a minor semidiurnal component with a speed of 27.968°/msh.

ν_2—a minor semidiurnal component with a speed of 28.513°/msh.

Tidal Phase is the lag time of appearance of high tide between the various locations experiencing the same tide; the phase is a reflection of wave celerity and is expressed in degrees relative to the duration of a full cycle represented by 360°; e.g., the duration of a full M2 cycle is about 12 h and 25 min corresponding to its speed of nearly 29° per hour (see Tidal Constituents).

Tidal Range is the elevation difference between successive high and low water levels at a specified location during the passage of a tide.

Tidal Regime describes the predictable behavioral characteristics of the tides under natural gravitational influence, i.e., not pertaining to meteorological disturbances.

Turbidity is a measure of the opacity of water from which the amount of suspended solids in the water may be inferred.

Units T (tera) = 10^{12} k (kilo) = 10^{3}
G (giga) = 10^{9} μ (micro) = 10^{-6}
M (mega) = 10^{6}

Wave Climate defines the behavioral height characteristics of waves under the influences of wind currents.

Wetland is an area of relatively poorly drained land with its characteristic ecosystem.

Wind Setup is the height or superelevation of a water body above its natural level caused by the impact of wind stress across its surface.

REFERENCES

Ackers, P. (1979). Tidal Power Projects—Australia. Proc. Colston Symposium No. 30, Tidal Power and Estuary Management, April 1978, Univ. of Bristol, Scientechnica, pp. 108–114.

Acres American Inc. (1981). *Preliminary Assessment of Cook Inlet Tidal Power, Phase I.* Report to Office of the Governor, State of Alaska (2 vols.). Columbia, Maryland, September.

Ad Hoc Working Party on Tidal Power (1978). *Tidal Power, The Next Step.* Report to Advisory Council on Research and Development for Fuel and Power, London.

Aisiks, E. G. (1992). Tidal Power in Argentina and Its Potential Development. In Proc. International Symposim on Tidal Power Stations, Sept.18–23. The Ministry of Fuel and Energy, Russian Federation, Moscow.

Amos, C. L. (1977). Effects of Tidal Power Structures on Sediment Transport and Loading in the Bay of Fundy–Gulf of Maine System. In Fundy Tidal Power and the Environment, Proc. *Workshop on Environmental Implications of Fundy Tidal Power*, Acadia University Institute, Wolfville, Nova Scotia, November 1976.

Amos, C. L. (1979). Sedimentation Resulting from Fundy Tidal Power. Proc. Colston Symposium (No. 30), Tidal Power and Estuary Management, April 1978, Univ. of Bristol, Scientechnica, pp. 173–179.

Amos, C. L. and K. W. Asprey (1979). Geophysical and Sedimentary Studies in the Chignecto Bay System, Bay of Fundy—A progress report. *Geological Survey of Canada Paper 79-IB*, pp. 245–252.

Amos, C. L., J. Zevenhuisen, and R. Beiko (1994). The Sediment Budget of Cumberland Basin, Chignecto Bay, Bay of Fundy, Natural Resources Canada, Internal Report, November.

Elements of Tidal-Electric Engineering. By Robert H. Clark

Atlantic Tidal Power Programming Board (1969). *Feasibility of Tidal Power Development in the Bay of Fundy*, Board Report and Committee Report. Ottawa. October.

Baker, A. C. (1991). *Tidal Power.* Peter Peregrinus, London.

Baker, G. C. (1983). The Annapolis Tidal Power Project. Paper presented at the annual general meeting Canadian Electrical Association, Halifax, NS, 26–28 June.

Baker, G. C. (1987). Environmental Impacts of Large-Scale Tidal Power Development in the Bay of Fundy. Paper prepared for Nova Scotia Tidal Power Corporation, May 1987.

Baker, G. C. (1992). The STRAFLO Turbine at Annapolis: Experience to Date. Proc. Int. Symposium on Tidal Power Stations, Murmansk, Sept. 18–23, 1991. Moscow.

Baker, G. C. and E. van Walsum (1983). Development Options for Fundy Tidal Power Potental. In Proc. of 7th Canadian National Energy Forum, Halifax, Nova Scotia, February. pp. 49–70. Sponsored by Canadian National Committee, World Energy Conference.

Baker, G. C. and E. van Walsum (1983). Development Options for Fundy's Tidal Power Potential. In Energy Developments in Atlantic Canada, 7th Canadian National Energy Forum, Halifax. Feb. 21 and 22, pp. 9–71.

Barreau, A. (1997). Communication introductive au colloque du 30e anniversaire de l'usine marémotrice de la Rance. *La Houille Blanche*, No. 3, p. 13.

Bay of Fundy Tidal Power Environmental Assessment Panel (1978). Draft Guidelines for Preparation of an Environmental Impact Statement for a Proposed Bay of Fundy Tidal Power Project. Ottawa, prepared for Tidal Power Review Board.

Bay of Fundy Tidal Power Review Board and Management Committee (1977). *Reassessment of Fundy Tidal Power.* November, Ottawa.

Bernstein, L. B. (1965). *Tidal Energy for Electric Power Plants* (translated from Russian). Israel Program for Scientific Translations, Jerusalem.

Bernstein, L. B. (1972). Kislaya Guba Experimental Tidal Power Plant and Problem of the Use of Tidal Energy. In *Tidal Power*, T. J. Gray and O. K. Gashus (Eds.). Proc. Int. Conf. on Utilization of Tidal Power. May 1970, Halifax, Nova Scotia. Plenum Press, New York, pp. 215–238.

Bernstein. L. B. (1974). Kislogubsk: A Small Station Generating Great Expectations. *Water Power*, May, pp. 172–177.

Bernstein, L. B. (1979). Time to Build Tidal Hydroelectric Plants. *Energo Export*, Vol. 8.

Bernstein, L. B. (1986). Tidal Power Engineering in the USSR. *Water Power and Dam Construction*. March, pp. 37–40.

Bernstein, L. B. (1992). Soviet TPP Projects (Mezen, Tugur, Kolskaya) and their realization. Proceedings International Symposium on Tidal Power Stations, Murmansk, 18–23 Sept. 1991, Moscow.

Bernstein, L. B. (1995). Tidal Power Development—A Realisitic, Justifiable and Topical Problem of Today. *IEEE Trans. Energy Conversion*, Vol. 10, No. 3, Sept., pp. 591–599.

Bernstein, L. B. (Ed.) (1996). *Tidal Power Plants* (English translation), E. M. Wilson and W. O. Song (Eds.). Korea Ocean Research and Development Institute, Seoul.

Bonnefille, R. (1978). Historical Sketch of the Studies of the French Chausey Islands Tidal Power Projects Plan. Proc. Int. Symposium on Korea Tidal Power, Korean Ocean Research and Development Institute. Seoul, Korea. Nov. 14–15.

Bosc, J. (1997). Les groupes bulbes de la Rance aprés trente ans d'exploitation—Retour d'expérience. *La Houille Blanche*, No. 3, pp. 18–24.

Braikevitch, M. (1970). Straight Flow Turbine. In *Tidal Power*, T. J. Gray and O. K. Gashus (Eds.). Proc. Conf. on Utilization of Tidal Power, May 1970, Halifax, Nova Scotia, Plenum Press, New York, pp. 415–434.

British Columbia Hydro and Power Authority (1979). *Report on Tidal Power in British Columbia.* Vancouver, June.

British Columbia Hydro and Power Authority (1984). *Queen Charlotte Islands Tidal Power Study, Juskatla Narrows Overview Report*, July.

Canadian Atlantic Power Group Ltd. (1988). Assessment of Retiming Tidal Power from Bay of Fundy Using Compressed Air Energy Storage. Prepared for Energy, Mines and Resources Canada, Ottawa.

Chamberlain, N. (1989). Nuclear Fuel: The Thinking Man's Alternative. Article based on 55th Melchett Lecture to the Institute of Energy at the Royal Institution, London. *Energy World*, April, pp. 11–15.

Charlier, R. H. (1969). Harnessing the Energies of the Oceans—A Review and Bibliography, Parts 1 and II. *MTS Journal*, Vol. 3, No. 3, pp. 13–32, and No. 4, pp. 59–81.

Charon, A. (1997). La conduite automatique de la centrale de la Rance. *La Houille Blanche*, No. 2, pp. 32–35.

Cheng, X. (1985). Tidal Power in China. *Water Power and Dam Construction*, February, pp. 33–36.

Clark, R. H. (1978). Reassessing the Feasibility of Fundy Tidal Power. *Water Power and Dam Construction*, June, pp. 35–41.

Clark, R. H. (1979). The Economics of Fundy Tidal Power. In *Proc. Int. Symposium on Wave and Tidal Energy.* Univ. of Kent, Canterbury, Sept. 1978. BHRA Fluid Engineering, Bedford, England.

Clark, R. H. (1997). Prospects for Fundy Tidal Power. *La Houille Blanche*, No. 3. pp. 79–85.

Clark, R. H. and A. N. Karas (1979). Studies of Tidal Power from the Bay of Fundy. In *Proc. Int. Conf. on Future Energy Concepts*, 30 Jan–1 Feb. IEE, London.

Collins, N. H. (1984). Potential Fish Mortality Associated with Large Hydroelectric Turbines. In *Update of the Marine Environmental Consequences of Tidal Power Development in the Upper Reaches of the Bay of Fundy*, Canadian Technical Report of Fisheries and Aquatic Sciences No. 1256, Bedford Institute of Oceanography, Dartmouth, Nova Scotia, June, pp. 551–564.

Conroy, R. D., H. R. Gibson, and D. Phillips (1992). *The Mersey Barrage—Mechanical and Electrical Equipment.* Proc. Conf. on Tidal Power, Institution of Civil Eng. London, 19–20 March.

Cotillon, J. (1979). La Rance Tidal Power Station—Review and Comments. Proc. Colston Symposium No. 30, *Tidal Power and Estuary Management*, April 1978, Univ. of Bristol, Scientechnica, pp. 49–56.

Crookshank, N. L. (1984). *Hybrid Modelling of Coastal Estuaries Using Array Processors.* Proc. ARRAY Conference, Denver, Colorado.

Daborn, G. R. (Ed.) (1977). Fundy Tidal Power and the Environment. *Proc. Workshop on Environmental Implications of Fundy Tidal Power.* Wolfville, Nova Scotia, Nov. 4 and 5, 1976, Acadia University Institute, Wolfville, Nova Scotia.

Dalton, F. K. (1961). Tidal Electric Power Generation. *Journal Royal Astronomical Society of Canada*, Vol. 55, Part 1, pp. 23–33, Part 2, pp. 57–72.

Dawson, W. B. (1920). *The Tides and Tidal Streams.* Dept. of Naval Services, Ottawa.

Dean, R. G. (1966). Tides and Harmonic Analysis, Chapter 4. In *Estuary and Coastline Hydrodynamics*, A. T. Ippen (Ed.). Engineering Societies Monographs, McGraw-Hill, New York.

de Carvalho, M. L. (1982). Possibilitades de Geraçao na Costa Maranhao-Para-Amapa (Tidal Power Generation Potential of the Maranhao-Para-Amapa Coast). Memorandum prepared for Centrais Eletricas Braxileiras S.A., ELETROBRAS, Rio de Janeiro, June 1982.

Decouer, (1890). Appareil Hydraulique avec nouveau modèle de turbine pour l'utilization continue de la force des marées. *Brevets*, Nos. 205–339, 29 avril.

Delory, R. P. (1986). *The Annapolis Tidal Generating Station.* Paper presented at Conf. on Water for Energy, 3rd Int. Symposium on Wave, Tidal, OTEC and Small Scale Hyro Energy sponsored by BHRA, Brighton, England, May.

Denton, A. A. (1978). Second Report on Marine Operations of F.I.P. Commission on Sea Structures. *F.I.P. Notes.*

Department of Mines and Resources, Canada (1945). *Report on Tidal Power, Petitcodiac and Memramcook Estuaries, Province of New Brunswick*, Ottawa.

Derrington, J. A. (1979). The Use of Concrete Caissons for River Barrages. Proc. Colston Symposium No. 30, *Tidal Power and Estuary Management*, April 1978, Univ. of Bristol, Scientehnica, pp. 133–140.

DeWolfe, D. L. (1977). Tidal Measurement Program of the Bay of Fundy—Gulf of Maine Tidal Regime. *Lighthouse*, Vol. 15, pp. 17–24.

Douma, A. and G. D. Stewart (1981). Annapolis Straflo Turbine Will Demonstrate Bay of Fundy Tidal Power Concept. *Modern Power Systems*, Vol. 1, No. 53(7).

Eckstein, O. (1961). A Survey of the Theory of Pubic Expenditure Criteria. In *Public Finances, Needs, Sources and Utilization*, Princeton University Press, Princeton, NJ.

Electrcité de France (1992). The Rance Tidal Power Station—A Quarter of a Century! Proc. Int. Symposium on Tidal Power Stations, Murmansk, Sept.18–23, 1991. Moscow.

EPRI (1987). Turbine-Related Fish Mortality: Review and Evaluation of Studies. AP-5480 Research Report 2694–4, Palo Alto, California.

Fentzloff, H. E. (1972). The Tidal Power Plant "San José" Argentine. In Tidal Power, T. J. Gray and O. K. Gashus (Eds.). *Proc. Conf. on Utilization of Tidal Power*, May, 1970, Halifax, Nova Scotia, Plenum Press, New York, pp. 257–266.

Fraser, D. C. (1979). Risk Minimization in Giant Projects. Proc. Int. Conference on The Successful Accomlishment of Giant Projects, 17–18 May 1979, London. Organized by OYEZ-IBC.

Funke, E. R. (1981). Hybrid Modelling as Applied to the Bay of Fundy. *Proc. 8th Canadian Congress of Applied Mechanics*, Moncton, June.

Funke, E. R., J. Ploeg, and N. L. Crookshank (1979). Modelling Requirements and Techniques for Tidal Power Developments in the Bay of Fundy. Hydraulics Laboratory Tech. Report LTR-HY-71, National Research Council, Ottawa.

Furst, G. B. and S. Sud (1978). Raw Tidal Energy Absorption Capability of a Power System. *Transactions of the IEEE on Power Apparatus and Systems*, Vol. PAS-97, No. 5, pp. 1910–1917.

Garrett, C. J. R. (1977). Tidal Influences on the Physical Oceanography of the Bay of Fundy and Gulf of Maine. In *Tidal Power and the Environment*, G. B. Daborn (Ed.). *Proc. of Workshop on the Environmental Implications of Fundy Tidal Power*, Wolfville, Nova Scotia, Nov. 4 and 5, Acadia University Institute Publication, No. 28.

Gates, A. D. and A. D. J. O'Neill (1977). The Climate of the Bay of Fundy. Proc. Workshop on Environmental Implications of Fundy Tidal Power, Wolfville, Nova Scotia, Nov. 4 and 5, 1976, Acadia University Institute, pp. 116–124.

Gibrat, R. (1966). *Energie des Marées*. Presses Universitaires de France, Paris.

Gibrat, R. (1976). The Current Revival in Tidal Power Studies. Paper delivered to ASTEO, Paris, France, March.

Godin, G. (1974). The Power Potential of Ungava Bay and Its Hinterland. *Water Power*, Vol. 25, pp. 167–171.

Godin, G. (1980). Cotidal Charts for Canada. Manuscript Report Series No. 55, Marine Sciences and Information Directorate, Dept. of Fisheries and Oceans, Ottawa.

Gordon, D. C. and M. J. Dadswell (Eds.) (1984). *Update on the Marine Environmental Consequences of Tidal Power Development in the Upper Reaches of the Bay of Fundy*. Proc. Workshop Organized by Fundy Environmental Studies Committee of APICS, 1982, Université de Moncton. Canadian Technical Report of Fisheries and Aquatic Sciences No. 1256.

Gordon, D. C. and A. R. Longhurst (1979). The Environmental Aspects of a Tidal Power Project in the Upper Reaches of the Bay of Fundy. *Marine Pollution Bulletin*, Vol. 10, pp. 38–45.

Grant, A. D. (1997). Simplified Tidal Barrage for Small-Scale Applications. *Journal Energy Engineering*, April, pp. 11–19.

Greenberg, D. A. (1976). *Mathematical Description of the Bay of Fundy–Gulf of Maine Numerical Model.* Tech. Note No. 16, Marine Environmental Data Service, Environment Canada, Ottawa.

Greenberg, D. A. (1977). Effects of the Tidal Power Development on the Physical Oceanography of the Bay of Fundy and Gulf of Maine. *Proc. Workshop on Environmental Implications of Fundy Tidal Power*, Wolfville, Nova Scotia, Nov. 4 and 5, 1976. Acadia University Institute. pp. 200–232.

Greenberg, D. A. and C. L. Amos (1983). Suspended Sediment Transport and Depositional Modelling in the Bay of Fundy. *Can. J. Fish. Aquatic Sci.*, Vol. 40 Suppl. 1, pp. 20–34.

Gwynn, J. D., F. Spargaren, and A. J. Woestenenk (1972). Modern Techniques for Barrage Construction. In *Tidal Power*, T.J. Gray and O. K. Gashus (Eds.). Proc. Int. Conf. on Utilization of Tidal Power, May 1970, Halifax, Nova Scotia. Plenum, New York, pp. 337–364.

Hadley, B. and L. E. Lindestrom (1990). The Sidney A. Murray, Jr. Hydro Power Project. Paper presented at a seminar of the Institute Mech. Engineers, London, April.

Harleman, D. R. F. (1966). Tidal Dynamics in Estuaries. In *Estuary and Coastline Hydrodynamics*, A. T. Ippen (Ed.). Engineering Societies Monographs, McGraw-Hill, New York, Chapter 10, Part II.

Haws, E. T., N. Reilly, and P. Wood (1992). *Prospective Tidal Power Projects in the Kimberley Region of Western Australia.* Paper presented at a conf. on tidal power, March 1992, Institution of Civil Engineers, London.

Haydock, J. L. and J. G. Warnock (1982). Compressed Air Energy Storage and Tidal Power, Conference on New Approaches to Tidal Power, Halifax, Nova Scotia.

Heaps, N. A. (1972). Tidal Effects Due to Water Power Generation in the Bristol Channel. In *Tidal Power*, T. J. Gray and O. K. Gashus (Eds.). Proc. Int. Conf. on Utilization of Tidal Power, May 1970, Halifax, Nova Scotia. Plenum, New York, pp. 435–456.

Herrmann, B. (1975). Air Storage Offers Peak Power for Northern Germany. *Energy International*, June, pp. 26–28.

Hydraulics Research Station (1980). River Scheldt Surge Barrier, Study of Estuary Sediments, Report No. Ex928.

Intergovernmental Panel on Climate Change (IPCC) (1990). *Scientific Assessment of Climate Change.* IPCC is sponsored jointly by WMO and UNEP.

International Commission on Large Dams (1980). Dam Projects and Environmental Success. Report prepared by the Committee on the Environment and approved by the 48th Executive Meeting, Rome, Oct.

International Joint Commission (1961). *Report on Investigation of the International Passamaquoddy Tidal Power Project.* Docket 72, Washington and Ottawa, April.

International Passamaquoddy Engineering Board (1959). *Investigation of the International Passamaquoddy Tidal Power Project.* Report to the International Joint Commission, Ottawa, Washington. October.

Ippen, A. T. (Ed.) (1966). *Estuary and Coastline Hydrodynamics*, Engineering Societies Monographs, McGraw-Hill, New York.

Ippen, A. T. and D. R. F. Harleman (1966). Tidal Dynamics in Estuaries. In *Estuary and Coastline Hydrodynamics*, A. T. Ippen (Ed.). Engineering Societies Monographs, McGraw-Hill, New York, Chapter 10.

Jellet, J. H. (1948). Breakwaters at Arromanches Harbour. *The Civil Engineer at War*, Vol. 2. Institution of Civil Engineers, London.

Karadi, G. M., R. J. Krizek, and S. C. Csallany (Eds.) (1971). Pumped Storage Development and Its Environmental Effects. In *Proc. Int. Conf. on Pumped Storage Development and Its Environmental Effects*, Sept. 19–24, 1971, Univ. of Wisconsin, Milwaukee, American Water Resources Assoc., Urbana, Illinois.

Karas, A. (1977). System Planning for Bay of Fundy Tidal Power Developments. *Transactions of IEEE on Power Aparatus and Systems*, Vol. PAS-97, No. 5, pp. 1600–1606.

Kestner, F. J. T. (1975). The Loose-Boundary Regime of the Wash, *Geographical Journal*, Vol. 141, pp. 369–412.

Keulegan, G. M. (1965). Model Laws for Coastal and Esturaine Models, In *Estuary and Coastline Hydrodynamics*, A. T. Ippen (Ed.). Engineering Socleties Monographs, McGraw-Hill, New York, Chapter 17.

Kirby, R. (1997). Environmental Consequences of Tidal Power in a Hydpertidal Muddy Regime: The Severn Estuary. *La Houille Blanche*, No. 3, pp. 50–56.

KORDI (1986). *Korea Tidal Power Study—1986.* Vol. 1 Data Report; Vol. 2 Technical Report. Prepared for Korea Electric Power Corp. Seoul, Korea BSPI 00050-124-2.

Koh, C-H, (1997). Korean Megatidal Environments and Tidal Power Projects: Korean Tidal Flats—Biology, Ecology and Land Uses by Reclamations and Other Feasibilities. *La Houille Blanche*, No. 3, pp. 66–78.

Laboratoire d'Hydraulique de France (1992). *Telemac-2D, Integrated Modelling System for Hydrodynamics and Transport in Natural Domains.* Laboratoire National d'Hydraulique of EDF-DER, Chatou, France.

Lambert, M. (1997). Bilan de la protection cathodique des ouvrages du barrage de la Rance—De la conception au résultat final. *La Houille Blanche*, Vol. 52, No. 3, pp. 36–41.

Lee, S. T. Y. and C. Deschamps (1977). Mathematical Model for Economic Evaluation of Tidal Power in the Bay of Fundy. *Transactions of IEEE Power Apparatus and Systems*, Vol. PAS-97, No. 5, pp. 1769–1778.

Leggett, W. C. (1977). Ocean Migration Rates of American Shad, *J. Fish. Res. Board Can.* Vol. 34, pp. 1422–1426.

Lewis, J. G. (1963). The Tidal Power Resources of the Kimberleys. *Journal Institution of Engineers*, Australia, Vol. 35, No. 12, pp. 333–345.

Longhurst, A. (1977). Overview Assessment of Draft Synthesis Report on Environmental Implications on Tidal Power Development in the Upper Bay of Fundy, Appendix F. Reassessment of Fundy Tidal Power, Management Committee, November, Ottawa.

Makela, G. A. (1983). Float-In Powerhouses, *Journal Energy Engineering*, ASCE, Vol. 109, No. 2, June, pp. 43–49.

Marphenin, N., O. I. Malutin, A. N. Pantulin, and N. M. Pertzova (1997). A Tidal Power Environmental Impact Assessment: Case Study of the Kislogubskaija Experimental Tidal Power Station (Russia). *La Houille Blanche*, No. 3, pp. 101–105.

Matthews, M. E. and R. M. Young (1992a). Feasibility of a Conway Barrage. *Proc. Conference on Tidal Power*, Institution of Civil Engineers, London, March.

Matthews, M. E. and R. M. Young (1992b). Tidal Energy from the Wyre. *Proc. Conference on Tidal Power*, Institution of Civil Engineers, London, 19–20 March.

Mauboussin, G. (1972). L'usine marémotrice de la Rance. In *Tidal Power*, T. J. Gray and O. K. Gashus (Eds.). Proc. of Int. Conference on the Utilization of Tidal Power, May 24–29, 1970, Nova Scotia Technical College, Halifax, Nova Scotia. Plenum, New York, pp. 189–214.

Mayers, P. (1980). Analysis of Lead Times and Approval Processes Facing Canadian Electrical Utilities. *Proc. of World Energy Conference*, Vol. 4B, p. 269.

McManus, J. (1979). Sediments in Estuaries. Proc. Colston Symposium No. 30, *Tidal Power and Estuary Management*, April 1978, Univ. of Bristol, Scientechnica, pp. 180–187.

Merlin, A., P. Sandrin, J. M. Grés, and M. Hilairet (1982). AGRA, the New Operation Model for the "La Rance" Tidal Power Plant. *Transactions of IEEE Power Apparatus and Systems*, Vol. PAS-101, No. 2, pp. 290–294.

Merrow, E. W., S. W. Chapel, and C. Worthing (1979). A Review of Cost Estimation in New Technologies. *Implications for Energy Process Plants.* Prepared by Rand

Corp. under contract R-248l-DOE for U.S. Department of Energy, Washington, D.C. July.

Miller, M. (1978). Choice of Hydro-Electric Equipment for Tidal Energy. *Proc. Int. Symposium on Korea Tidal Power*, Korea Ocean Research and Development Institute, Seoul, Nov. 14–15, pp. 179–230.

Modern Power Systems (1981). Energy Short Chinese Turn to Hydro, Conservation, Nuclear and Tidal Power. Vol. 1, No. 5, May, pp. 21–27.

Monaghan, J. J. (1994). Simulating Free Surface Flows with SPH. *Journal of Computational Physics*, Vol. 110, pp. 399–406.

Mosonyi, E. (1987). *Water Power Development, Vol. 1, Low Head Power Plants*, Akadémiai Kiadó, Budapest.

Muirhead, S. J. (1992). The Environmental Effects of Tidal Energy, *Proc. Tidal Power Conference*, Institution of Civil Engineers, Mar. 19–20, London.

Munday, J. C., T. T. Afoldi, and C. L. Amos (1980). Application of a System for Automated Multidate Landsat Measurement of Suspended Sediment. *Water Quality Bulletin*, Vol. 5, No. 1, pp. 6–10.

National Council on Electric Energy (1994). Tidal Power—A Position Paper prepared by the National Committee on Electric Energy. College of Electrical Engineers, The Institution of Engineers, Australia, April 20th.

Nekrasov, A. V. and D. A. Romanenkov (1997). On Effects Produced by Tidal Power Plants upon Environmental Conditions in Adjacent Sea Areas, *La Houille Blanche*, No. 3, pp. 88–95.

Nessim, M., D. Attwood, and V. Swail (1991). *The Analysis of Combined Environmental Extremes for Offshore Structures*, Vol. III of Proc. First Int. Offshore and Polar Eng. Conference, Aug. 11–16, Edinburgh, UK, sponsored by Int. Society of Offshore and Polar Engineers.

Nova Scotia Tidal Power Corporation (1982). Fundy Tidal Power, Update '82. Halifax.

Nova Scotia Tidal Power Corporation (1985). Fundy Tidal Power—Stage 1. Halifax, December.

Potts, R. and E. Wilson (1992). The Mersey Barrage—Further Assessment of the Energy Yield, *Conf. on Tidal Power*, Institution of Civil Engineers, London, March.

Prandle, D., E. R. Funke, N. L. Crookshank, and R. Renner (1980). The Use of Array Processors for Numerical Modelling of Tidal Estuary Dynamics. *Proc. 17th Conf. on Coastal Engineering*, Sidney, Australia.

Rétière, Ch. (1984). Conséquences écologiques de la construction et du fonctionment de l'usine marémotrice de la Rance, *La Houille Blanche*, No. 8, pp. 584–585.

Rétière, C., C. Bonnot-Courtois, P. LeMao, and N. Desroy (1997). Etat écologique du Bassin maritime de la Rance au terme de 30 ans de fonctionnement de l'usine marémotrice. *La Houille Blanche*, No. 3, pp. 106–107.

Rice, R. G. and G. C. Baker (1987). The Annapolis Experience. In *Proc. OCEANS '87: The Ocean—An International Workplace*, Vol. 2, pp. 391–396.

Rice, R. G. and G. C. Baker (1992). Annapolis: The Straflo Turbine and Other Operating Experiences. *Proceedings of Conference on Tidal Power: Trends and Developments*, March, Institution of Civil Engineers, Thomas Telford Ltd., London.

Salvi, J. (1997). Usine marémotrice de la Rance: Reconstruction des alternateurs. *La Houille Blanche*, No. 3. pp. 42–46.

Saunders, D. W. (1974). Kimberley Tidal Power Revisited. Institution of Engineers, Australia. Conference on Energy Resources for Electrical Power Generation, Melbourne. National Conference Publication No. 74/4, pp. 47–55.

Saville, T. (1958). Wave Run-up on Shore Structures. *Trans ASCE*, Vol. 123, pp. 139–150.

Semple, J. R. (1984). Possible Effects of Tidal Power Development in the Upper Bay of Fundy on Anadromous Fish Passage. In *Update on the Marine Environmental Consequences of Tidal Power Development in the Upper Reaches of the Bay of Fundy*, Canadian Technical Report of Fisheries and Aquatic Sciences, No. 1256, Bedford Institute of Oceanography, Halifax, Nova Scotia, June, pp. 565–576.

Seoni, R. M. (1977). Major Electrical Equipment Proposed for Tidal Power Plants in the Bay of Fundy. *Transactions of IEEE Power Apparatus and Systems*, Vol. PAS-98, No. 5, pp. 1750–1760.

Serrer, M. and N. L. Crookshank (1995). HYDA: An Integrated Hydronumerical Modelling Environment. Proceedings of the 12th Canadian Hydrotechnical Conference, CSCE, Ottawa, pp. 201–209.

Severn Barrage Committee (1981a). *Tidal Power from the Severn Estuary*, Vols. 1 and 2, Energy Paper No. 46. Department of Energy, HMSO, London.

Severn Barrage Committee (1981b). *Sensitivity and Risk Analysis.* Report prepared for the Committee by the Project Management Group, Dept. of Civil Engineering, Univ. of Manchester Institute of Science and Technology, December 1980.

Severn, R. T., D. Dineley, and L. E. Hawker (Eds.) (1979). *Tidal Power and Estuary Management.* Proc. Colston Research Society Symposium No. 30, April 16–19, 1978, University of Bristol, Scientechnica.

Severn Tidal Power Group (1986). *Tidal Power from the Severn.* Interim Study Report, Vol. 1 (report), Vol. 2 (engineering/economic studies—Cardiff Weston scheme and English Stones scheme), STPA, London.

Severn Tidal Power Group (1989). *The Severn Barrage Project: General Report.* Energy Paper Number 57, Department of Energy, HMSO, London.

Sharma, H. R. (1986). Kachchh Tidal Power Project. In *Water for Energy. 3rd Int. Symp. on Wave, Tidal, OTEC and Small Scale Hydro Energy*, Brighton, England, May, pp. 179–190.

Sharma, H. R. (1989). Personal communication.

Shaw, T. L. (Ed.) (1980). *An Environmental Appraisal of Tidal Power Stations; with Particular Reference to the Severn Barrage.* Pitman Publishing, London.

Shaw, T. L. (1997). Study of Tidal Power Projects in the UK, with the Exception of the Severn Barrage. *La Houille Blanche*, No. 3, pp. 57–65.

SHP News (1985). *Newsletter of the Asia-Pacific Regional Network for Small Hydro Power*, No. 2. Hangzhou, China.

SHP News (1997). Exploitation of Tidal Energy in China. *Newsletter of the Asia-Pacific Regional Network for Small Hydro Power*. Hangzhou, China, Spring.

Sogreah (1959). *Usina maremotriz del Golfo de San Jose, Anteproyecto.* Societé Grenobloise d'Etudes et d'Applications Hydrauliques, Grenoble, France.

Sogreah (1981). *Garolim Tidal Power Plant.* Summary report on feasibility studies prepared for Korea Electric Co. Seoul, Korea.

Song, W. O. (1992). A Review of Korean Tidal Power Studies. *Proc. Int. Symposium on Tidal Power Stations*, Sept. 1991, Murmansk, pp. 65–72.

Subrahmanyan, K. S. (1978). Tidal Power in India. *Water Power and Dam Construction*, June, pp. 42–44.

Swales, M. C. and E. M. Wilson (1978). Optimization of Tidal Power Generation. *Water Power*, April, Vol. 20, pp. 109–114.

Sykes, A. (1979). The Project Overview. In *Sucessfully Accomplishing Giant Projects.* Proc. of an International Conference on the Successful Accomplishment of Giant Projects, A. Sykes (Ed.), London, May 1978, organized by OYEZ-IBC.

Thomann, G. (1978). Environmental Aspects in Modern Pumped Storage Planning. *Water International*, Sept., pp. 8–10.

Tidal Power Corporation (1982). *Fundy Tidal Power Update '82.* Halifax, Nova Scotia.

Tidal Power Corporation (1985). *Fundy Tidal Power, Stage 1.* Halifax, Nova Scotia. December.

United Nations (1975). *Report on Tidal Power in India.* Dept. of Technical Cooperation for Development, New York.

U.S. Army Coastal Engineering Research Center (1973). *Shore Protection Manual Vol. 2*, Washington, D.C.

Walia, M., R. German, and A. Z. Ahmed (1979). Overwater Drilling in Extreme Tidal Conditions. Paper presented at Marine Geotechnical Engineering Conference, Calgary, Alberta, April.

Waller, D. H. (1972). Enviromental Aspects of Tidal Power Development. In *Tidal Power*, T. J. Gray and O. K. Gashus (Eds.). Proc. Int. Conf. on Utilization of Tidal Power, May 1970, Halifax, Nova Scotia. Plenum, New York, pp. 611–625.

Wang, Y. (1991). Interaction of Large Square Caissons in Random Waves. Vol. III of Proc. of First Int. Offshore and Polar Eng. Conf., Edinburgh, UK, Aug. 11–16, sponsored by Int. Society of Offshore and Polar Engineers.

Warnock, J. G. (1971). Trends Established from 20 years of Modern Pumped Storage Practice and for the Future. In *Proc. of the Int. Conference on Pumped Storage Development and Its Environmental Effects*, University of Wisconsin, Milwaukee American Water Resources Assoc., Urbana, Illinois, pp. 9–32.

Warnock, J. G. (1978). Organization and Management for Tidal Power Studies—II. *Proc. Int. Symposium on Korea Tidal Power*, Korea Ocean Research and Development Institute, Seoul, Nov., pp. 111–118.

Warnock, J. G. (1979). A Giant Project Accomplished—Design Risk and Engineering Management. *Proc. Int. Conf. on the Successful Accomplishment of Giant Projects*, May, London. Organized by OYEZ-IBC.

Warnock, J. G. (1987). Energy from the Oceans—A Review of Recent and Proposed Tidal Power Installations. Presented to World Federation of Engineering Organizations Technical Congress. Vancouver, British Columbia, May 26th.

Warnock, J. G. and R. H. Clark (1992). The State of the Art in Tidal Power Recovery. In *Ocean Energy Recovery: The State of the Art*, R. J. Seymour (Ed.). Amer. Soc. Civil Eng., New York.

Warnock, J. G. and B. L. Smith (1984). Trends in Pumped-Storage Development, *Water Power & Dam Construction*, August, pp. 15–19.

Warrick, R. and H. Oerlamans (1990). Sea Level Rise, *Climate Change: The IPCC Scientific Assessment*, pp. 257–281, Cambridge University Press, Cambridge, UK.

Wayne, W. W. (1977). *Final Report on Tidal Power Study for the United States Energy Research and Development Administration*. Contract No. E(49-18)-2293 with Stone and Webster Engineering Corporation, Boston.

Westerink, J. J. and W. G. Gray (1991). Progress in Surface Water Modelling. *Reviews of Geophysics*, Supplement, pp. 210–217, American Geophysical Union, April.

Whipple, W. Jr. (1963). Problems of National Water Resources Planning. Paper delivered at 57th Meeting of Princeton University, May 23–24.

Wilson, E. (1979). Tidal Power Projects in Canada, India and Korea. Proc. Colston Symposium No. 30, *Tidal Power and Estuary Management*, April 1978, Univ. of Bristol, Scientechnica, pp. 94–107.

Wilson, E. (1992). *The Mersey Barrage Project.* Proc. Int. Symposium on Tidal Power Stations, Sept 18–23, 1991. The Ministry of Fuel and Energy, Russian Federation, Moscow.

Zhu, X. (1980). Personal comnunication, July 14. Ministry of Water Conservancy, Beijing, China.

BIBLIOGRAPHY

Moyse, C. M. (1978). *Bay of Fundy Environmental and Tidal Power Bibliography.* Fisheries and Marine Service Technical Report No. 822, Department of Fisheries and Environment, Ottawa, Canada.

Plant, S. (1985). *Bay of Fundy Environmental and Tidal Power Bibliography*, 2nd ed. Technical Report No. 1339, Fisheries and Aquatic Sciences, Bedford Institute of Oceanography, Dartmouth, Nova Scotia.

Song, Won-Oh and Ki-Sup Song (Eds.) (1993). *Tidal Power Bibliography,* 3rd ed. Korean Ocean Research and Development Institute, Seoul, Korea. (Contains 1936 items of reference.)

Elements of Tidal-Electric Engineering. By Robert H. Clark

INDEX

Elements of Tidal-Electric Engineering. By Robert H. Clark